AF289872

Current Topics in Microbiology 130 and Immunology

Peptides as Immunogens

Edited by
H. Koprowski and F. Melchers

With 21 Figures

Springer-Verlag
Berlin Heidelberg New York
London Paris Tokyo

Professor Dr. Hilary Koprowski
The Wistar Institute
36th Street at Spruce
Philadelphia, PA 19104
USA

Professor Dr. Fritz Melchers
Institute for Immunology
Grenzacherstr. 487
CH-4005 Basel

ISBN-13:978-3-642-71442-9   e-ISBN-13:978-3-642-71440-5
DOI: 10.1007/978-3-642-71440-5

Library of Congress Catalog Card Number 15-12910

2123/3130-543210

Preface

The humoral response of the immune system to a foreign antigen usually requires the recognition of two antigenic determinants. The one, called the carrier, is recognized by T-lymphocytes, the other, called the hapten, by B-lymphocytes. As a consequence, T- and B-lymphocytes proliferate, B-lymphocytes produce hapten-specific antibodies, and the system develops memory to the antigens. It was long thought that antigens would form a bridge to mediate the cooperation of T- and B-lymphocytes. However, it now appears that antigens are broken down to fragments which then act as carrier determinants for T-lymphocytes.

The cells which originally process antigen are called antigen-presenting cells. They have phagocytic properties. They can take up and degrade antigens, in the case of proteins to peptides.

The peptides of protein antigens reappear on the surface of the antigen-presenting cells, where they must become associated with membrane proteins encoded by genes of the major histocompatibility complex (MHC) in order to be recognized by T-lymphocytes. To activate helper T-lymphocytes which cooperate in antibody responses, MHC class II molecules have to be expressed on the surface of the antigen-presenting cells. Once T-lymphocytes have become activated, they are ready to cooperate with B cells.

Antigen-presenting cells are macrophages but can also be B-lymphocytes. In the latter case, a protein antigen can be taken up through binding of one of its heptenic determinants to the hapten-specific antibody displayed on the surface of the B cells. It is then processed and a peptide presented on the surface of the B cells as a carrier determinant in conjunction with class II MHC encoded membrane proteins. T-lymphocytes specific for this peptide-MHC class II complex can, then, recognize the B cell, cooperate with it, and induce it for hapten-specific antibody secretion. The remarkable consequence of this view of T-cell-dependent activation of antibody production by B cells is the realization that processed antigens, e.g., peptides, are separate physical entities of antigens that have become specifically separate from haptens. Peptides should, therefore, be able to act as carrier determinants which can be used to selectively prime T-lymphocyte compartments of the immune system without stimulating B-lymphocytes.

This selective priming of parts of the immune system is likely to offer new practical approaches for immunizations and, thus, for the development of vaccines.

FRITZ MELCHERS
HILARY KOPROWSKI

Table of Contents

Indexed in Current Contents

List of Contributors

ARNON, R., Department of Chemical Immunology, Weizmann Institute of Science, P.O.B. 26, Rehovot 76100, Israel

BERKOWER, I., Metabolism Branch and Laboratory of Mathematical Biology, National Cancer Institute, National Institutes of Health, Bethesda, MD 20892, USA

BERZOFSKY, J.A., Metabolism Branch and Laboratory of Mathematical Biology, National Cancer Institute, National Institutes of Health, Bethesda, MD 20892, USA

CEASE, K., Metabolism Branch and Laboratory of Mathematical Biology, National Cancer Institute, National Institutes of Health, Bethesda, MD 20892, USA

CORNETTE, J., Metabolism Branch and Laboratory of Mathematical Biology, National Cancer Institute, National Institutes of Health, Bethesda, MD 20892, USA

DELISI, C., Metabolism Branch and Laboratory of Mathematical Biology, National Cancer Institute, National Institutes of Health, Bethesda, MD 20892, USA

DIETZSCHOLD, B., The Wistar Institute, 36th Street at Spruce, Philadelphia, PA 19104, USA

DYRBERG, T., Department of Immunology, Research Institute of Scripps Clinic, La Jolla, CA 92037, USA

GAMMON, G.M., Department of Microbiology, University of California, Los Angeles, CA 90024, USA

HEBER-KATZ, E., The Wistar Institute, 36th Street at Spruce, Philadelphia, PA 19104, USA

LANZAVECCHIA, A., The Basel Institute for Immunology, Grenzacherstr. 487, CH-4005 Basel

MARGALIT, H., Metabolism Branch and Laboratory of Mathematical Biology, National Cancer Institute, National Institutes of Health, Bethesda, MD 20892, USA

OLDSTONE, M.B.A., Department of Immunology, Research Institute of Scripps Clinic, La Jolla, CA 92037, USA

SCHWARTZ, R.H., Laboratory of Immunology National Institute of Allergy and Infectious Diseases, National Institutes of Health, Bethesda, MD 20892, USA

SERCARZ, E.E., Department of Microbiology, University of California, Los Angeles, CA 90024, USA

Peptides as Immunogens:
Prospects for Synthetic Vaccines

R. Arnon

1 Introduction

The basis for the utilization of synthetic peptides as immunogens was laid during our early studies on the antigenicity of proteins. The initial findings that covalent attachment of tyrosine oligopeptides to gelatin by polymeric techniques resulted in augmentation of its immunogenicity (Sela and Arnon 1960) led us to the synthesis of both linear and branched polymers of amino acids, capable of initiating an immune response (Sela et al. 1962). The availability of these synthetic antigens permitted a systematic elucidation of the molecular basis of antigenicity, including the role of such variables as chemical composition, size, shape, accessibility of epitopes, electrical charge, optical configuration, and mainly spatial conformation in rendering a molecule immunogenic and in dictating its antigenic specificity (Sela 1969).

In 1971 we first showed that it is possible to synthesize a peptide which elicits antibodies capable of recognizing a native protein (Arnon et al. 1971). The epitope in this case was the "loop" of hen egg-white lysozyme, demonstrated previously to be a conformation-dependent immunodominant region of the protein (Arnon and Sela 1969). As a corollary, we suggested that by adequate molecular engineering it should be possible to design synthetic materials that would elicit antiviral immunity, leading eventually to multivalent synthetic vaccines (Arnon 1972).

The rapid development of DNA cloning and sequencing techniques which facilitate protein sequencing has made the synthetic approach more feasible and practical than before. Primary sequences are available today for many proteins, enabling the synthesis of any selected region, using either chemical

Department of Chemical Immunology, Weizmann Institute of Science, Rehovot, Israel

Current Topics in Microbiology and Immunology, Vol. 130

laboratory techniques, which are advisable in the case of short peptides, or genetic engineering procedures in the case of longer protein segments, subunits, or even whole proteins. Moreover, the synthetic approach should allow the choice of carrier as well as adjuvant, the goal being the production of multivalent vaccines with built-in adjuvanticity.

2   Identification of Antigenic Determinants

Macromolecular antigens usually express a large number of possible antigenic determinants, or epitopes, that dictate their antigenic specificity. However, only a limited number of the potential antigenic sites are important for immunogenicity, namely, immunodominant ones. In the case of an antigen that possesses biological activity, even fewer epitopes are involved in the neutralizing immune capacity.

It is useful to distinguish between two types of epitopes – *sequential* and *conformational* (SELA 1969). Whereas a sequential determinant is viewed as a sequence of several residues in its unfolded form, a conformational determinant is defined by a number of residues that are maintained in a particular conformation, to which long-range interactions of the levels of secondary, tertiary, and quaternary structure contribute as well. A parallel distinction introduced by ATASSI and SMITH (1978) differentiates between *continuous* and *discontinuous* determinants. A continuous determinant is defined as a sequence of residues exposed at the surface of a native protein, whereas a discontinuous one consists of the juxtaposition in space of residues that are not adjacent in the primary structure.

When one is dealing with protein antigens, it seems that the parameters which are the most influential in contributing to the immunological properties are spatial conformation and accessibility. Accessibility is crucial since the interaction of any epitope with either the antibodies or the specific receptors on the immunocompetent cells necessitates its exposure on the antigenic molecule. As a result, antigenic determinants were identified in many native proteins as segments that are positioned at "corners" of the folded polypeptide chain and fully or partially exposed on the surface of the protein molecule (ATASSI 1975). Hence, the probability that a particular region contains surface residues provides one way of predicting the antigenic structure.

Spatial conformation has long been known to play a decisive role in determining antigenic specificity. There is cumulative evidence that a drastic change in antigenic properties occurs upon denaturation of native proteins or unfolding of their polypeptide chains (BENJAMINI et al. 1972; ARNON 1974; CRUMPTON 1974), or even as a result of more subtle conformational alterations (CRUMPTON and WILKINSON 1966). This has led to the conclusion that most of the antigenic determinants of proteins are conformational determinants. Antibodies specific toward such determinants will not necessarily react with isolated peptides derived from the molecule. It is, therefore, of interest that many investigators have been successful in raising antipeptide polyclonal antisera that recognize native protein (e.g., ARNON et al. 1976; LERNER et al. 1981; GREEN et al. 1981). This high rate of success may partly be attributed to the fact that the relative flexibility of peptides in solution can be mimicked by local disorder in short segments of the protein. Indeed, it was shown recently that the majority of continuous epitopes of tobacco mosaic virus (TMV) protein, myoglobin, lysozymes, and myohemerythrin correspond to surface regions in those proteins that possess a high segmental mobility (WESTHOF et al. 1984; TRAINER et al. 1984). It is possible that segmental motion in proteins provides the necessary degree of conformational similarity with isolated peptides and that this ensures sufficient antigenic cross-reactivity between the intact molecule and an excised fragment. This is true, however, mainly for relatively short peptide segments. When longer regions are concerned, they usually consist of a more rigid structure and the right conformation is essential for the specificity (WESTHOF et al. 1984; SHAPIRA et al. 1985).

Identification of antigenic determinants can be made by fragmentation of the native protein, either by chemical cleavage or controlled proteolysis. The resultant fragments are then screened for immunologically active components that can bind to antibodies and interfere with their interaction with the intact antigen. Data obtained by those methods have resulted in the identification of antigenic determinants in several protein antigens such as TMV protein (BENJAMINI et al. 1964; ALTSCHUH et al. 1983), sperm whale myoglobin (ATASSI 1975) or hen eggwhite lysozyme (SHINKA et al. 1962; ARNON and SELA 1969). Alternatively, monoclonal antibodies with neutralizing activity could be useful in identifying relevant antigenic sites. This approach is particularly useful for identification of determinants involved in some biological phenomena, e.g., virus neutralization. An illustrated example is the localization of the various antigenic sites on influenza hemagglutinin (HA) (WILEY et al. 1981), in which monoclonal antibodies with or without neutralizing capacity served as an indispensable tool. Crystallographic studies of the three-dimensional structure of proteins serve as an additional source of information in regard to antigenic structure. For most proteins, however, although the primary structure is known, crystallographic data are not yet available. Hence, various methods based on computational analysis of hydrophilicity (ROSE 1978), high antigenicity (HOPP and WOODS 1981), or flexibility (KARPLUS and SCHULTZ 1985) of various regions in the molecule have been suggested for the prediction of the more likely antigenic determinants. Such cumulative information on segments in protein molecules with the right "ingredients" might be suggestive as regards the design of synthetic peptides that will serve as immunogens.

A completely novel and original approach for identifying epitopes that cross-react with particular antigens was recently suggested by GEYSSEN and colleagues. The strategy they employed for identifying sequential determinants in a protein involves a systematic synthesis of all its possible heptapeptide units and measurement of their reactivity with antibodies against the native protein. The peptides showing the highest reactivity presumably contain sequential epitopes (GEYSSEN et al. 1984). Alternatively, if the starting point is a monoclonal antibody that is highly reactive with the protein, it is possible to identify by systematic synthesis of all possible short peptides the one with the highest binding properties. Such a peptide may bear little or even no direct relation to the original epitope but it mimics its ability to bind specifically to the antibody and is called a "mimetope" (GEYSSEN et al. 1985).

3 Protein-Specific Synthetic Peptide Antigens

Once antigenic determinants in a protein have been identified, it is possible to synthesize them chemically for investigation of their immunological properties. This approach has now been employed in the case of many protein antigens and has clearly demonstrated that synthetic peptides can induce antibodies that react with the intact protein in its native form. In several cases the spatial conformation was shown to play an important role in this activity, as for example in the case of hen egg white lysozyme. A fragment of this protein, consisting of residues 60–83 with a single disulfide bond and called a "loop", was shown to be immunologically active. This region, which is exposed on the surface of the molecule, is a conformation-dependent antigenic determinant (MARON et al. 1971; PECHT et al. 1971). A conjugate of a synthetic peptide corresponding to this sequence elicited antibodies with identical specificity that were reactive with native lysozyme and still recognized a conformational determinant (ARNON et al. 1971).

Another example is the coat protein of MS-2 coliphage. A CNBr fragment of the molecule P_2, consisting of residues 88–108, was identified as an antigenic region which was capable of inhibiting the process of neutralization of the phage by antiphage antibodies. A conjugate, P_2-A–L, containing a synthetic peptide corresponding to the same sequence, induced antibodies reactive with the intact protein and capable of neutralizing the viability of the MS-2 phage (LANGBEHEIM et al. 1976). Furthermore, when the synthetic adjuvant MDP (CHEDID et al. 1978) was covalently linked to P_2-A–L, the resulting conjugate elicited in rabbits very efficient MS-2–neutralizing antibodies even when administered in physiological medium, thus providing a completely synthetic antigen with built-in adjuvanticity that leads to antiviral response (ARNON et al. 1980). These findings paved the way to the study of synthetic peptide vaccines. In our laboratory the efforts in this direction concentrated mainly on two systems – influenza virus and cholera toxin (CT) – as will be described herein.

4 Synthetic Peptides Leading to Anti-Influenza Immune Response

The influenza virus provided a suitable model for studying the synthetic approach to vaccination, since detailed information is available on its structure and function, as well as its genetic variations and the serological specificities of its various strains. Influenza hemagglutinin (HA) is the major antigenic protein against which neutralizing antibodies are directed, and it is also responsible for the attachment of the virus to cells (LAVER and VALENTINE 1969). Both the amino acid sequence (LAVER et al. 1981) and the three-dimensional structure (WILSON et al. 1981) of this protein are known, and data on its immunochemical properties (JACKSON et al. 1979) and the location of its antigenic sites (WILEY et al. 1981) are also available. Another advantage of this system is that once immunogenic fragments of it have been synthesized, the immune response they elicit can be assessed on four different levels: (a) The immunochemical reactivity, namely, the capacity of the antibodies to cross-react with the entire protein; (b) their inhibitory effect on HA biological activity; (c) the in vitro neutralization of virus plaque formation in tissue-cultured cell monolayers; and (d) the in vivo protection of mice against challenge infection.

The first peptide we synthesized, before the three-dimensional structure was known, consisted of 18 amino acid residues corresponding to the sequence 91–108 of the influenza HA molecule. This region, which is common to at least 12 H3 strains, is part of a larger cyanogen bromide fragment, previously shown by JACKSON and his colleagues (1979) to be immunologically active. According to our computer-predicted folded structure of the HA polypeptide chain, this peptide segment should have comprised a folded region with a short alpha-helical section, and hence an exposed area in the molecular structure. It also contains three tyrosine and two proline residues that had been demonstrated in our previous studies to play a dominant role in the antigenicity of several proteins or synthetic antigens. The 91–108 peptide that contains these "ingredients" was anticipated to be immunologically reactive.

Indeed, a conjugate of this peptide with tetanus toxoid elicited in both rabbits and mice antibodies that reacted immunochemically with the synthetic peptide, as well as with the intact influenza virus of several strains of type A. These antibodies were capable of inhibiting the capacity of the HA of the relevant strains to agglutinate chicken red blood cells. They also interfered with the in vitro growth of the virus in tissue culture, causing a reduction of up to 60% in viral plaque formation. But, most importantly, as shown in Table 1, mice immunized with the peptide-toxoid conjugate were partially protected against further challenge infection with the virus (MULLER et al. 1982).

Table 1. Protection of mice against challenge infection with influenza virus

Infectious strain	Group	Incidence of Infection at 10^{-2} Dilution into egg	Lung virus titer (EID[c])
A/Tex/77	Immunized[a]	19/36 (52%)	$10^{-1.98}$
	Control[b]	21/23 (91%)	$10^{-3.56}$
A/PC/75	Immunized	4/18 (22%)	$10^{-0.61}$
	Control	8/19 (42%)	$10^{-1.53}$
A/Eng/42/72	Immunized	8/18 (44%)	$10^{-1.61}$
	Control	15/21 (71%)	$10^{-2.9}$
A/PR/34(H1)	Immunized	18/19 (95%)	$10^{-3.47}$
	Control	20/21 (95%)	$10^{-3.95}$

[a] Immunized with a conjugate of the peptide 91-108 and tetanus toxoid in FCA
[b] Control groups were injected with the tetanus toxoid alone in FCA
[c] Egg infective dose

One of the crucial factors concerning the influenza vaccine is the tremendous genetic variations among the various virus strains – shifts and drifts – and their reflection on the serological specificities (LAVER and AIR 1979). The four proposed antigenic sites of the HA constitute variable regions in the molecule (WILSON et al. 1981). The peptide 91-108 was deliberately chosen to be part of a conserved region, and hence the protective effect it elicits is manifested with several strains of type A influenza virus (Table 1). This finding may serve as an indication that the synthetic approach might lead to multivalent vaccines for cross-strain protection.

Incidentally, tyrosine 98 is one of the residues that form the tentative receptor site of the molecule (WILSON et al. 1981). Although not designated by the authors as one of the antigenic sites, the peptide 91-108 could readily be visualized as an exposed region that has an immunological imprint, particularly in the infective form of the virus. This corroborates the explanation for the protective effect it elicits. It is even possible that in the intact virus this region forms a hidden determinant, but when present on a synthetic antigen, it might lead to the formation of antibodies that can reach the antigenic zone in situ and inactivate the infective virus.

As is evident from the three-dimensional structure of the influenza HA (WILEY et al. 1981), the region 140-146, which forms antigenic site A, is a "loop" of seven amino acid residues unusually protruding from the surface of the molecule. It is also exposed in the trimeric structure assumed by the HA in the spikes on the virus surface. One would expect that peptides corresponding to this region where "natural" immunogenic determinants of the HA molecule are located would elicit a better and more specific immune response against the virus. It was of interest, therefore, to synthesize peptides covering the loop region for immunization purposes. JACKSON et al. (1982) used a synthetic peptide representing the sequence 123-151, which includes the loop region, for immunization of rabbits. They did not find significant binding between the anti-peptide antibodies and the X-31 virus. However, antibodies raised against the intact X-31 virus did exhibit some binding to the synthetic peptide.

In our studies (SHAPIRA et al. 1984), four synthetic peptides were examined. Two of them corresponded to the sequence 139-146, with either Gly or Asp at position 144. The third peptide corresponded to the sequence 147-164, and the fourth included both regions and corresponded to the sequence 138-164. All four peptides, conjugated to tetanus toxoid, elicited high antibody titer against the respective homologous peptides, with a significant degree of cross-reactivity among them. Of particular interest was the high cross-reactivity between the two octapeptides differing at position 144 (Gly to Asp exchange). This finding contrasts with the significant effect of the same exchange on the serological specificity of the intact HA (LAVER et al. 1981), which is manifested in the existence of monoclonal antibody-selected variants that escape neutralization owing to this single amino acid substitution.

The four synthetic peptides differed in their cross-reactivity with the intact H3 influenza virus. Thus, antibodies induced by the two longer peptides, 147-164 and 138-164, showed significant binding with the intact virus, whereas the extent of binding of the antisera to the two short octapeptides 139-146 was essentially insignificant. In contrast, antibodies raised against the intact virus A/Mem/102/72 or against the isolated HA were capable of recognizing the synthetic "loop" octapeptides but did not react at all with the region 147-164.

The role of the size of the peptide fragment in this system was emphasized even more in the interference of the antibodies with the biological activity of the virus, in which only the longer fragment 138-164 proved effective. Its reactivity was manifested both in the capacity of the antiserum to inhibit the HA activity of the intact virus and in the fact that active immunization of mice resulted in their partial protection against infection with the A/Eng/42/72 strain, with which the sequence of the synthetic peptide corresponds.

These results indicate that the loop region 140-146, although constituting a major naturally occurring antigenic determinant of the intact virus, is too short to fold into a loop, but that, when forming a part of the longer peptide 138-164, folding to the right conformation is facilitated. These findings are illustrative of the importance of both the length and the conformation of a synthetic peptide for its immunological reactivity.

5 CT and Coli Toxin

Detoxified bacterial toxins or toxoids have been used for vaccination, e.g., against tetanus and diphtheria, for almost a hundred years. Hence, this family of proteins seems of value as a model for investigating the synthetic approach for vaccination. Indeed, AUDIBERT et al. (1981) have shown that active immunization against diphtheria can be achieved by a synthetic tetradecapeptide consisting of residues 188-201 in the amino acid sequence of the diphtheria toxin. Conjugates of this peptide, or the hexadecapeptide 186-201, linked covalently to a protein carrier or a synthetic copolymer, elicited antibodies in guinea pigs that not only bind specifically to the toxin but also neutralize its dermonecrotic activity and lethal effect. The same results were achieved if the conjugate contained in addition the synthetic adjuvant muranyl dipeptide attached to the carrier to replace the regularly employed Freund's adjuvant. In this case a completely synthetic soluble molecule elicited antitoxin activity (AUDIBERT et al. 1982).

We have used the synthetic approach to elicit antibodies cross-reactive with cholera toxin (CT) and *E. coli* toxin that are capable of partially neutralizing their biological activity. The toxin of *Vibrio cholerae* is composed of two subunits, A and B. Subunit A activates adenylate cyclase, which triggers the biological activity, whereas subunit B is responsible for binding to cell receptors and expresses most of the immunopotent determinants. Antibodies to the B subunit are capable of neutralizing the biological activity of the intact toxin; hence, the B subunit was the obvious candidate for the pursuit of synthetic peptide immunogens. Moreover, in view of the high level of sequence homology between the B subunits of CT and the heat-labile toxin of *E. coli* (LT), we have also investigated whether the same synthetic peptides will also cross-react with *E. coli* toxin.

```
                                               10                                            20
 [Thr]-Pro-Gln-[Asn]-Ile-Thr-[Asp]-Leu-Cys-[Ala]-Glu-Tyr-[His]-Asn-Thr-Gln-Ile-[His]-Thr-[Leu]-
 [Ala] Pro  Gln [Thr] Ile  Thr [Glu] Leu  Cys [Ser] Glu Tyr [Arg] Asn Thr Gln Ile [Tyr] Thr [Ile]

                                               30                                            40
 [Asn]-Asn-Lys-Ile-[Phe]-Ser-Tyr-Thr-Glu-Ser-[Leu]-Ala-Gly-Lys-Arg-Glu-[Met]-Ala-Ile-[Ile]-
 [Asp] Asn  Lys Ile [Leu] Ser  Tyr  Thr  Glu  Ser [Met] Ala  Gly  Lys  Arg  Glu [Val] Ala  Ile  Ile

                                               50                                            60
 Thr-Phe-[Lys]-[Asn]-Gly-[Ala]-Thr-Phe-Glu-Val-Glu-Val-Pro-Gly-Ser-Gln-His-Ile-Asp-Ser-
 Thr  Phe [Met] [Ser] Gly [Glu] Thr  Phe  Glu  Val  Glu  Val  Pro  Gly  Ser  Gln  His  Ile  Asp  Ser

                                               70                                            80
 Gln-Lys-Lys-Ala-Ile-Glu-Arg-Met-Lys-[Asn]-Thr-Leu-Arg-Ile-[Ala]-Tyr-Leu-Thr-Glu-[Ala]-
 Gln  Lys  Lys  Ala  Ile  Glu  Arg  Met  Lys [Asp] Thr  Leu  Arg  Ile [Thr] Tyr  Leu  Thr  Glu [Thr]

                                               90                                           100
 Lys-[Val]-[Glu]-Lys-Leu-Cys-Val-Trp-Asn-Asn-Lys-Thr-Pro-[His]-[Ala]-Ile-Ala-Ala-Ile-Ser-
 Lys [Ile] [Asp] Lys  Leu  Cys  Val  Trp  Asn  Asn Lys  Thr  Pro [Asn] [Ser] Ile  Ala  Ala  Ile  Ser

                  103
 Met-[Ala]-Asn
 Met [Lys] Asn
```

Fig. 1. Amino acid sequence of the B subunits of cholera toxin (*upper line*) and the heat-labile toxin of *E. coli* (*lower line*). Residues in the *boxes* are those that differ in the two toxins

Six peptides corresponding to sequences 8-20, 30-42, 50-64, 69-85, 75-85, and 83-97 of the CT B subunit (Fig. 1), and designated CTP 1 to CTP 6 respectively were synthesized, coupled to tetanus toxoid, and used for immunization of rabbits and mice (JACOB et al. 1983). Evaluation of the immunological reactivity of the resultant antisera by radioimmunoassay, immunoprecipitation and immunoblotting, indicated that each of the antisera reacted with the respective homologous peptide, but four antisera out of the six also reacted, to different extents, with the intact B subunit and the native CT. The antisera against the peptides CTP 1 and CTP 6 showed a very high level of reactivity with the respective homologous peptides but reacted only slightly (several order of magnitude difference) with the intact B subunit of CT. Incidentally, anti-CTP 1 also reacted with the A subunit of CT. On the other hand, peptide CTP 3 induced antibodies which, though of lower absolute titer, gave a very strong cross-reactivity with the intact toxin, very similar in its level to that of the homologous peptide-antipeptide reaction. Furthermore, this peptide was the only one that reacted with antiserum against the native CT.

Antibodies to CTP 2 (12 residues) and CTP 5 (11 residues) were reactive with the respective homologous peptides but did not react at all with the intact protein. Elongation of each of them by four or five amino acid residues resulted in peptides that induced antibodies cross-reactive with the intact B subunit and the holotoxin. However, elongation of a peptide does not always result in augmentation of reactivity; thus, for example, elongation of CTP 3 (15 residues) by 5 amino acid residues led to a lowering of the cross-reactive capacity of the elicited antibodies (JACOB et al. 1986). These findings indicate that the capacity of a peptide to transconform to fit the structure of the native protein is dependent on its size and may be prevented either by the peptide segment's being too short to possess the correct conformation or too long and already possessing a stabilized conformation which is different from that of the protein.

Of most interest among these peptides were CTP 1 (residues 8-20) and CTP 3 (residues 50-64). Antisera against these two peptides exerted significant inhibition of the biological activity of CT. The toxic effect of CT can be demonstrated by skin vascular permeation and fluid accumulation in ligated small intestinal loops, as well as at the biochemical level, by the induction of adenylate cyclase. The inhibitory effect of the antipeptide sera was manifested in all the assays of the biological activity of the toxin with very good correlation between the biochemical level and the end biological effect of the toxin. In both cases the inhibition reached a value of approximately 60% (JACOB et al. 1983, 1984a).

As mentioned above and also demonstrated in Fig. 1, there is a high level of sequence homology between the B subunits of CT and the heat-labile toxin of *E. coli* (LT). Moreover, an immunological relationship was demonstrated between the two toxins, with the existence of both shared and specific antigenic determinants (LINDHOM et al. 1983). Since the toxin of pathogenic strains of *E. coli* is the causative agent of diarrhea in many tropical countries and, owing to its wide spread, probably presents a more serious health problem than cholera, it was of interest to investigate whether the synthetic peptides derived from CT might cross-react with the LT and/or provide a comparable degree of protection against the heterologous toxin.

Indeed, we have demonstrated that the antiserum elicited by CTP 3 (residues 50-64) is highly cross-reactive with the LT in both radioimmunoassay and immunoblotting (JACOB et al. 1984b). This is not surprising, since in this region the sequence homology between the two toxins is complete (Fig. 1). The antiserum against CTP 1 (residues 8-20) was also cross-reactive with the two toxins, though to a much lesser extent. However, antisera to both CTP 1 and CTP 3, which are inhibitory toward CT, were found equally effective in neutralizing the biological activity of the *E. coli* LT. This was manifested by the significant inhibition of both adenylate cyclase induction and the fluid secretion into ligated ileal loops of rats. The inhibition by the anti-CTP 3 was expected in view of the high level of immunological cross-reactivity. As for the anti-CTP 1, its high efficacy in the inhibition of the biological activity, contrasting with the very low serological activity, may imply that the inhibition it confers is due to an interaction with subunit A and not necessarily with subunit B, in spite of the lack of homologous sequence in the A subunit of the toxins. The immunological relationship in this case could stem from similarities in conformation rather than sequence.

Another interesting phenomenon observed with the synthetic peptides of CT was their capacity to "prime" the experimental animal toward the intact toxin. Thus, rabbits primed by a single administration of the synthetic peptide conjugates, which as such does not lead to an immune response, and subsequently boosted by a subimmunogenic dose (1 µg) of CT, too small to give any effect on its own, demonstrated a significant level of anti-CT immune response, including toxin-neutralizing capacity (JACOB et al. 1986b). The priming effect was achieved irrespectively of whether or not the immunization with the peptide conjugate itself could lead to a neutralizing response. In fact, even conjugates of peptides that do not induce any CT cross-reactive antibodies at all, such as CTP 2, led to a primary effect, though of a lower level. A similar priming effect was observed previously with synthetic peptides of polio virus (EMINI et al. 1983), indicating that this may be a general phenomenon.

The ability of synthetic peptides to prime the immune system may be of general value. The basis for this effect probably lies in the heterogenicity of the immune response. The peptide elicits antibodies of various specificities, and a secondary stimulus with the native protein serves as a selective booster only to those cells which produce antibodies cross-reacting with it. In the case of the peptide CTP 3, priming was achieved to both cholera toxin and the toxin of *E. coli* of either human or porcine source (H-LT or P-LT), an observation that might have a practical value for vaccination.

6 Synthetic Peptide Vaccines with Built-In Adjuvanticity

The choice of adjuvant for synthetic materials is crucial, since these may often be of relatively low immunogenicity. In most animal experimentations, Freund's complete adjuvant (FCA) is used for augmenting the immune reactivity. This adjuvant, which consists of a water-in-oil emulsion containing killed mycobacteria, is a very effective adjuvant eliciting high-level and long-lasting immunity. However, it is not suitable for human use, since it induces local reactions and granulomas, as well as inflammation and fever, on account of both the low metabolism of mineral oil and the mycobacteria. Efforts are being made to replace the FCA by other adjuvants, preferably by eliminating the mineral oil and/or replacing the mycobacteria by less damaging materials.

It has been demonstrated that the mycobacteria in Freund's adjuvant can be replaced by less damaging substances. The minimal structure that can substitute for mycobacteria was identified as *N*-acetylmuramyl-L-alanyl-D-isoglutamine, designated MDP for muramyl dipeptide (ELLOUZ et al. 1974). This material and some of its synthetic analogs have already been used in combination with synthetic antigens (ARNON et al. 1980; AUDIBERT et al. 1982; CARELLI et al. 1982). Recently, MDP has been employed in combination with the 91-108 peptide of HA for induction of anti-influenza response, and with the CTP 3 peptide of CT.

In the case of the influenza peptide, MDP was similar to FCA in the induction of antibodies specific toward the peptide. As for the cross-reactivity with the intact virus, although a marked difference was observed between the titer of antibodies raised in the presence of FCA and MDP, both adjuvants were capable of inducing protective immunity against a viral challenge (SHAPIRA et al. 1985). This implies a probable participation of cellular immunity induced by the synthetic peptide and is in accord with several recent publications which emphasize the role of cellular immunity in antiviral protection (ADA et al. 1981). Incidentally, MDP was efficient only when coupled covalently to the conjugate (91-108)TT, but in that form it led to the highest protection rate against in vivo challenge, even higher than that induced in the presence of FCA.

Similar results were obtained with the CTP 3 peptide of CT. Its conjugates with the synthetic polymer (T,G)-A–L elicited protective anticholera antibodies when administered in FCA. A similar protective response was obtained if MDP was covalently attached to the latter immunogen (JACOB et al., unpublished results). The completely synthetic MDP-CTP 3 (T,G)-A–L, when administered in aqueous physiological medium, elicited in rabbits antibodies with CT-neutralizing capacity and thus constitutes a synthetic vaccine with built-in adjuvanticity.

7 Concluding Remarks

The results presented above, as well as the cumulative data from other studies with synthetic peptide vaccines (summarized in a recent review, ARNON 1985), illustrate the tremendous potential of synthetic peptides as immunogens and in the preparation of some vaccines. This approach, if materialized, will offfer several advantages: (a) It could allow the attachment of several peptides to the same carrier molecule, thus providing a multivalent synthetic vaccine; (b) it might eliminate the formation of antibodies against many irrelevant antigens that contaminate the essential immunogen and thereby reduce undesired effects; (c) It might permit the production of vaccines with built-in adjuvanticity that would be efficient when administered in physiological medium and would thus eliminate the hazard inherent in most of the presently available adjuvants. The exact methodology for predicting the most efficient peptide for each system needs further investigation and improvement. However, when finally identified, synthetic peptides and their conjugates may provide the immunogens of choice for future vaccines.

References

Ada GL, Leung KN, Ertl H (1981) Imm Rev 58:5
Altschuh D, Hartman D, Reinbolt J, Van Regenmortel MHV (1983) Mol Immunol 20:271
Arnon R (1972) In: Kohen A, Klingberg AM (eds) Immunity of viral and rickettsial diseases. Plenum Press, New York, pp 209
Arnon R (1974) In: Lotan N, Goodman M, Blout ER (eds) Peptides, polypeptides and proteins. Wiley 1974. pp 538
Arnon R (1985) In: Van Regenmortel MHV, Neurath AR (eds) Immunochemistry of viruses: the basis for serodiagnosis and vaccines. Elsevier, New York, pp 139
Arnon R, Sela M (1969) Proc Natl Acad Sci USA 62:163
Arnon R, Maron E, Sela M, Anfinsen CB (1971) Proc Natl Acad Sci USA 68:1450
Arnon R, Bustin M, Calef E, Chaitchik S, Haimovich J, Novik N, Sela M (1976) Proc Soc Natl Acad Sci USA 73:2123
Arnon R, Sela M, Parant M, Chedid L (1980) Proc Natl Acad Sci USA 77:6769–6772
Atasi MZ (1975) Immunochemistry 12:423
Atassi MZ, Smith JA (1978) Immunochemistry 15:609
Audibert F, Jolivet M, Chedid L, Alouf JE, Boquet P, Rivaille P, Shiffert O (1981) Nature 289:593
Audibert F, Jolivet M, Chedid L, Arnon R, Sela M (1982) Proc Natl Acad Sci USA 79:5052
Benjamini E (1977) In: Attasi MZ (ed) Immunochemistry of Proteins, vol 2. Plenum Press, New York, pp 265

Benjamini E, Young JD, Simizu M, Leung CY (1964) Biochemistry 3:1115
Carelli C, Audibert F, Faillard J, Chedid L (1982) Intl J Immunopharmacol 4:290
Chedid L, Audibert F, Johnson AG (1978) Progr Allergy 25:63–105
Crumpton MJ (1974) In: Sela M (ed) The antigens, vol 2. Academic Press, New York, pp 1
Crumpton MJ, Wilkinson JM (1966) Biochem J 100:223
Ellouz F, Adam A, Ciorbaru R, Lederer E (1974) Biochem Biophys Res Comm 55:1317
Emini EA, Jameson BA, Wimmer E (1983) Nature 304:699
Geyssen HM, Meloen RH, Basteling SJ (1984) Proc Natl Acad Sci USA 81:3998
Geyssen HM, Rodda SJ, Mason TJ (1986) In: Porter R, Whelan J (eds) Synthetic peptides as antigens, Ciba Foundation Symposium, Pitman, London (in press)
Green N, Alexander H, Olson A, Alexander S, Shinnick TM, Sutcliffe JG, Lerner RA (1982) Cell 28:477
Hopp TP, Woods KR (1981) Proc Natl Acad Sci 78:3824
Jackson DC, Brown LE, White DO, Dopheide TAA, Ward CW (1979) J Immunol 123:2610
Jackson DC, Murray JM, White DO, Fagan CN, Tregear GW (1982) Virology 120:272
Jacob CO, Sela M, Arnon R (1983) Proc Natl Acad Sci USA 80:7611
Jacob CO, Pines M, Arnon R (1984a) EMBO J 3:2889
Jacob CO, Sela M, Pines M, Hurwitz S, Arnon R (1984b) Proc Natl Acad Sci USA 81:7893
Jacob CO, Harari I, Arnon R, Sela M (1986a) Vaccine 4:95
Jacob CO, Grossfeld S, Sela M, Arnon R (1986b) (in press)
Karplus PA, Schulz GE (1985) Naturwissenschaften 72:212
Langbeheim H, Arnon R, Sela M (1976) Proc Natl Acad Sci USA 73:4636
Laver GW, Air GM (eds) (1979) Structure and variations in influenza virus. Elesevier, North Holland
Laver WG, Valentine RC (1969) Virology 38:105
Laver WG, Air GM, Dopheide TA, Ward CW (1980) Nature 283:454
Laver WG, Air GM, Webster RG (1981) J Mol Biol 145:339
Lerner RA, Green N, Alexander H, Liu FT, Sutcliffe GJ, Shinnick TM (1981) Proc Nat Acad Sci USA 78:3403
Maron E, Shiozawa C, Arnon R, Sela M (1971) Biochemistry 10:763
Müller GM, Shapira M, Arnon R (1982) Proc Natl Acad Sci USA 79:569
Pecht I, Maron E, Arnon R, Sela M (1971) Eur J Biochem 19:369
Rose GD (1978) Nature 272:586
Sela M (1969) Science 166:1365
Sela M, Arnon R (1960) Biochim Biophys Acta 40:382
Sela M, Fuchs S, Arnon R (1962) Biochem J 85:323
Sela M, Schechter B, Schechter I, Borek F (1967) Cold Spring Harbor, Symp on Quant Biol 32, 537
Shapira M, Arnon R (1985) Mol Immunol 22:23–28
Shapira M, Jolivert M, Arnon R (1986) Intl J Immunopharmacol (in press)
Shinka S, Imanishi M, Kuwahara O, Fujio H, Amano T (1962) Biken, J 5, 181
Trainer JA, Getzoff ED, Alexander H, Houghten RA, Olson AJ, Lerner RA, Hendrickson WA (1984) Nature 312:127
Westhof E, Altschuh D, Moras D, Bloomer AC, Mondragon A, Klug A, Van Regenmortel MHV (1984) Nature 311:123
Wiley DC, Wilson IA, Skehel JJ (1981) Nature 289:373
Wilson IA, Skehel JJ, Wiley DC (1981) Nature 289:366

Molecular Features of Class II MHC-Restricted T-Cell Recognition of Protein and Peptide Antigens: The Importance of Amphipathic Structures

J.A. Berzofsky, J. Cornette, H. Margalit, I. Berkower, K. Cease, and C. DeLisi

Much work has been done on the use of peptides as immunogens, but most has concentrated on antibody production (reviewed in Lerner 1984; Arnon 1984) rather than T-cell responses. There are certain potential problems with their use for antibody production that do not apply to T cells. Antibodies raised against peptides as immunogens generally cross-react with the native protein from which the peptides were derived with affinities several orders of magnitude lower than those for the peptide (reviewed in Berzofsky 1985a). Conversely, antibodies raised against the native protein generally cross-react with peptides derived from it with a lower affinity, although exceptions have been noted (Lando and Reichlin 1982). In both directions, cross-reactions seem to be most easily measured when the peptide corresponds to a more mobile portion of the protein (Tainer et al. 1984, 1985; Westhof et al. 1984); perhaps such portions are better able to share some conformational states with the peptide so that one can achieve an induced fit with antibodies made against the other (Berzofsky 1985a). In addition, there are antibodies made against the native protein that react with assembled topographic determinants consisting of amino acid residues far apart in the primary sequence that are brought together on the surface of the native molecule as it folds in its native conformation (Benjamin et al. 1984; Berzofsky 1985a). Such antigenic sites exist only on the surface topography of the native protein, and thus do not exist in any peptide segment of the protein. Studies of monoclonal antibodies to sperm whale myoglobin by ourselves and others are consistent with these ideas. Several monoclonal antibodies were found to react with sites assembled from residues far apart in the primary sequence, and these antibodies did not cross-react even with low affinity with any of the large cyanogen bromide (CNBr) cleavage fragments which together span the entire sequence of the protein (Berzofsky et al. 1980, 1982; East et al. 1982). Others have found similar monoclonal antibodies for other proteins (reviewed in Benjamin et al. 1984). At least for myoglobin, such antibodies appear to represent 30%–40% of serum antibodies raised against the native protein. Exhaustive depletion, by affinity chromatography, of antibodies binding to any of the three CNBr fragments of myoglobin left 28%–41% of the antibodies in each antiserum tested which still bound with high affinity in a radioimmunoassay to the native protein, but did not react detectably with any of the peptides spanning the sequence (Lando et al. 1982). This major

Metabolism Branch and Laboratory of Mathematical Biology, National Cancer Institute, National Institutes of Health, Bethesda, MD 20892, USA population of antibodies would not be elicited by immunization with any peptide segment of the protein. Such antibodies to assembled topographic sites may represent a large fraction of the neutralizing antibodies to viral proteins such as influenza hemagglutinin (WILEY et al. 1981; STAUDT and GERHARD 1983). These considerations have frequently limited the usefulness of short synthetic peptides as vaccines to elicit neutralizing antibodies.

In contrast, it was found over a quarter century ago that lymphocytes did not distinguish between native and denatured forms of a protein antigen (GELL and BENACERRAF 1959). This observation has been confirmed and extended many times in a large number of laboratories since then (reviewed in BERZOFSKY 1980), using both denatured proteins and peptides in comparison with native. This property, which probably can be attributed to the proteolytic antigen processing that occurs before helper T cells can recognize antigen "presented" on the surface of another cell (UNANUE 1984; BERZOFSKY 1985b, 1986), may make T-cell immunity a more achievable goal for synthetic vaccines.

T cells elicited by peptides are more likely than antibodies elicited by peptides to cross-react with the native protein, provided the peptide corresponds to a T-cell antigenic site in the native protein. Therefore, it would be useful to be able to predict such sites. This same property, the lack of recognition of tertiary structure (overall folding of the protein) by T cells, in contrast to antibodies, may make it easier to predict, from primary sequence, sites which stimulate T cells than to predict sites that bind antibodies, as local secondary structure is more easily predicted than tertiary. The remainder of this paper will summarize our recent studies of the structure of antigenic sites of myoglobin recognized by murine T cells, using myoglobin sequence variants and natural and synthetic peptides (BERKOWER et al. 1982, 1984, 1985, 1986; CEASE et al. 1986). These have led to the observation that such sites tend to be amphipathic structures (DELISI and BERZOFSKY 1985). We will examine and extend the theoretical analysis which appears to fit well with the limited data base of published sites recognized by T cells. This approach may be very useful for predicting sites recognized by T cells and thus, for example, for vaccine development, but a number of cautions on such applications must also be considered (see below).

Using myoglobin sequence variants from over 15 species, we observed that the majority of myoglobin-immune T cells of two high responder mouse strains responded to a site around residue 109 (BERKOWER et al. 1982, 1984). The immunodominance of this site depended on the *H-2* type of the mouse strain immunized, as the 109 region was immunodominant for $H\text{-}2^s$ and $H\text{-}2^d$ mice, whereas T cells from congenic mice of the $H\text{-}2^{i5}$ haplotype displayed a completely different pattern of cross-reactivity. Thus, *Ir* genes controlled immunodominance. Moreover, sperm whale myoglobin with Glu 109 elicited a different population of T cells from those elicited by horse myoglobin with Asp 109, as the former failed to cross-react with Asp 109-containing myoglobins and the latter failed to cross-react with Glu 109-containing myoglobins. Nevertheless, the 109 region was immunodominant in both. Therefore, the *Ir* gene control of immunodominance depended on some feature related to but independent of the T-cell specificity – for instance, possibly a nearby site that interacted with Ia (class II major histocompatibility [MHC]) antigens. This immunodominance of one or a few sites contrasts with the antibody response to myoglobin, in which almost the entire accessible surface can react (TODD et al. 1982; BENJAMIN et al. 1984; BERZOFSKY 1985a; DOROW et al. 1985). Perhaps this difference is related to the fact that T cells recognize antigen only in association with cell surface I a molecules, whereas antibodies will bind free antigen in solution.

To explore the mechanism behind this difference, and to further analyze the site around Glu 109 as well as other possible sites, we needed monoclonal populations of T cells. Therefore, long-term myoglobin-specific T-cell lines were prepared from H-2^d mice by the method of KIMOTO and FATHMAN (1980), and cloned in soft agar and subcloned by limiting dilution (BERKOWER et al. 1984, 1985). Five of the clones had the immunodominant cross-reactivity pattern indicative of Glu 109. These failed to react with the N-terminal and C-terminal CNBr cleavage fragments 1–55 and 132–153, but those tested did react with the middle fragment 56–131. (The specificity for the region around Glu 109 was subsequently confirmed with synthetic peptides, see below; CEASE et al. 1986.) However, nine other T-cell clones had a different cross-reactivity pattern which pointed to Lys 140. These clones cross-reacted with all myoglobins bearing a Lys at position 140, and only with those myoglobins. Moreover, they responded to the 22-residue CNBr cleavage fragment 132–153 containing Lys 140. Goosebeaked whale myoglobin has all the substitutions relative to sperm whale in the 132–153 region that horse myoglobin has except Lys → Asn at position 140; yet goosebeaked whale myoglobin stimulated these clones as well as sperm whale, whereas horse myoglobin failed to stimulate at all. This contrast confirms the assignment to 140 (BERKOWER 1985). (Again, synthetic peptides below will further confirm the mapping.)

However, the specificity of helper T cells is defined not solely by the protein epitope recognized, but also by the class II MHC molecule with which it is recognized. These were determined using H-2 recombinant and F_1 hybrid mice as sources of antigen presenting cells, and confirmed by blocking proliferation with monoclonal antibodies to I-A^d or I-E^d (BERKOWER et al. 1985). Strikingly, all of the clones specific for the site around Glu 109 were restricted to I-A^d, whereas all those specific for the site around Lys 140 were restricted to I-E^d. There were no exceptions to this correlation. Further, when a new uncloned myoglobin-specific T-cell line was stimulated with presenting cells bearing only I-A^d, only I-E^d (transfected L cells, a kind gift of R. Germain), or both, the Lys 140 site was found to be immunodominant when myoglobin was presented with I-E^d, whereas the Glu 109 site was immunodominant with I-A^d (BERKOWER et al. 1985). This correlation suggested a causal relationship, such as but not limited to "determinant selection" (ROSENTHAL 1978; BENACERRAF 1978) at the clonal level.

The finding that certain peptide epitopes are seen only in association with certain I a molecules has implications for synthetic peptide vaccines in an outbred species like man. Any given peptide may be recognized by only a subset of the population. Therefore, mixtures of peptides may be required for an effective synthetic vaccine.

In order to understand the molecular basis for this striking relationship between epitope and class II MHC restriction molecule, we examined these two immunodominant antigenic sites using cleavage and synthetic peptides. The hope was to see whether there were two subsites within each, one to interact with the T-cell receptor and a second to interact with the I a molecule to account for the preferential recognition with *I-A* ^{d} or *I-E* ^{d}. We started with the Lys 140 site as this was present on the smallest and most soluble CNBr cleavage fragment. This 22-residue fragment, 132–153, was further cleaved at Glu 136 or at Tyr 146 and the Lys 140-containing fragments 137–153 and 132–146 purified by HPLC. The former lost all activity, whereas the latter retained full activity. Thus, something in the stretch 132–136 was necessary, whereas 147–153 was unnecessary. The full activity of 132–146 was confirmed by solid-phase peptide synthesis and the series of synthetic peptides 133–146, 134–146, 135–146, 136–146, 137–146, 132–145, and 132–144 was also prepared (BERKOWER et al. 1986). All the activity was retained in the 11-residue peptide 136–146 (and the larger ones containing this), but 137–146 had lost over 90% of the activity. Thus, Glu 136 was identified as an important residue. Lys 133 also contributed to potency. The stimulatory activity of 132–145, but not 132–144, identified Lys 145 as another critical residue. These are spaced about every turn of the helix, so 133, 136, 140, and 145 are all on the same hydrophilic side of the native helix, perhaps in one site (BERKOWER et al. 1986). The 11-residue peptide must contain all the information necessary both for binding the T-cell receptor and also for interacting with I a, if such interaction occurs. This α-helix is amphipathic, with one side hydrophilic and the other hydrophobic. A suggestion that a second site, possibly for I a binding, may be on the hydrophobic side came from the following study of antigen processing.

Proteins must be presented to T cells by a metabolically active presenting cell such as a macrophage. Having a T-cell clone which saw the same site on the native protein and a 22-residue fragment (132–153), we asked whether both forms of the antigen required the same steps for presentation. We found that the lysosomal inhibitors chloroquine and NH_4Cl as well as the competitive protease inhibitor leupeptin inhibited presentation of the native molecule but not that of the fragment (STREICHER et al. 1984). Thus, the native protein requires some lysosomal proteolytic step not required by the peptide. Since the fragment differs from native protein in both size and conformation, we also used S-methyl myoglobin, an unfolded form of the intact protein, to distinguish these. Surprisingly, the unfolded form behaved like the fragment, demonstrating that conformation, not size, is the critical difference between the fragment and native protein in determining the requirement for processing (STREICHER et al. 1984). Since the main overall difference between native and the other forms is that the hydrophobic residues are buried in the native protein and exposed in the peptide and unfolded form, we suggest that the purpose of the processing may be to expose critical residues, possibly hydrophobic, necessary for interaction with I a or with the lipid membrane of the presenting cell. Examination of the structure of 132–146 in the native protein reveals a stretch of hydrophobic residues along one side of the helix which may represent this site.

Recently, the other immunodominant site around Glu 109 has been further delimited by synthetic peptides to be within the region 106–118. This peptide has at least as much activity as the native protein (CEASE et al. 1986). This region is also an amphipathic α-helix in the native protein.

In general, we suggest that amphipathicity (the property of having one face hydrophobic and one hydrophilic) may be an important property for peptides which can stimulate T cells (BERZOFSKY 1985b; DELISI and BERZOFSKY 1985), perhaps so that the hydrophobic face can interact with Ia or another structure on the surface of the presenting cell, while the hydrophilic face may interact with the T-cell receptor (e.g., Lys 133, 140, 145, and Glu 136 in our case). Studies of a lysozyme T-cell epitope by ALLEN et al. (1984) and of an ovalbumin T-cell epitope by WATTS et al. (1985) support the same concept. The results of GODFREY et al. (1984) for presentation of tyrosine-p-azobenzenearsonate and its analogues to T cells are also consistent with this hypothesis. In addition, theoretical calculations (PINCUS et al. 1983) have predicted and experiments (SCHWARTZ et al. 1985) have supported the importance of α-helical secondary structure in the major T-cell epitope of pigeon cytochrome c.

In general, the recognition site on a protein consists of chemical groups that are localized spatially, but well separated along the chain of the backbone. Under these circumstances, and without detailed structural knowledge of the folded chain, identification of those groups belonging to a given site is difficult. For the identification of T-cell antigenic sites, however, lack of geometric information on the native antigen turns out to be not as serious a problem as one might expect, and it is considerably less serious than for the analogous B-cell problem. The reason for this involves the antigenic processing requirement, that is, the requirement that T cells be stimulated not by the native protein, but by a proteolytic fragment which is presented to the T-cell receptor on the surface of a second cell.

The approach we have taken, then, is to examine amino acid sequences for a periodic variation in hydrophobic character, such that if these fold into an ordered periodic structure such as a helix or β-strand, all the hydrophobic side chains will be on one side and all the hydrophilic ones on the other, i.e., the structure will be amphipathic (DELISI and BERZOFSKY 1985). We assign a hydrophobicity value to each residue in the protein according to one of the published hydrophobicity scales (KYTE and DOOLITTLE 1982; ROSE et al. 1985) and search the resulting sequence of numbers for intervals in which the hydrophobicity values fluctuate either with a periodicity of approximately 3.6 residues per cycle (corresponding to an α-helix) or with a periodicity of approximately 2 residues per cycle (corresponding to a β-strand). The following mathematical analysis is used to detect such periodicity.

Standard techniques for detecting periodic variation in sequences lie in the methods of time series analysis and the discrete Fourier transform. Those methods are most useful, however, for study of periodicity in number sequences that are longer than the number of residues in a periodic structure in a protein, and must be used with some caution in this context. Let $h_1, h_2, \ldots, h_n$ denote the complete sequence of hydrophobicity values for a peptide or protein of n residues and let $h_k, h_{k+1}, \ldots, h_{k+l-1}$ or $\{h_{k+j}\}_{j=0}^{l-1}$ denote the hydrophobicity values for a segment of peptide from residue k to residue $k+l-1$. The sequence $\{h_{k+j}\}_{j=0}^{l-1}$ may be fluctuating about a mean different from 0, and our first step is to substract $\bar{h}_k = \left[\sum_{j=0}^{l-1} h_{k+j} \right] / l$ from each term. To detect periodic variation in the resulting sequence $\{h_{k+j}-\bar{h}_k\}_{j=0}^{l-1}$ we compare that sequence with a harmonic sequence

$$\cos\,(0\cdot\theta+\varphi),\quad \cos\,(1\cdot\theta+\varphi),\,\ldots,\,\cos\,[(l-1)\theta+\varphi] = \{\cos\,(j\theta+\varphi)\}_{j=0}^{l-1}$$

that has a known periodicity of $360°/\theta$ (the phase shift φ is adjusted to give the best match between the two sequences for the chosen frequency θ).

A close match between

$$\{h_{k+j}-\bar{h}_k\}_{j=0}^{l-1}\quad \text{and}\quad \{\cos\,(j\theta+\varphi)\}_{j=0}^{l-1}$$

for θ approximately 100°, say, signals a periodicity of approximately $360°/100°=3.6$ residues per cycle, and suggests that if the protein segment from residue k to residue $k+l-1$ were isolated from the protein, the segment would tend to form an amphipathic α-helix in an amphipathic environment (DELISI et al. 1986).

The comparison between the numbers $\{h_{k+j}-\bar{h}_k\}_{j=0}^{l-1}$ and the harmonic sequence $\{\cos\,(j\theta+\varphi)\}_{j=0}^{l-1}$ may be made in one of two ways, referred to respectively as inner product and least squares:

Inner Product. Compute the inner product $R(\theta)$ between the balanced data $\{h_{k+j}-\bar{h}_k\}_{j=0}^{l-1}$ and $\cos\{(j\theta+\varphi)\}_{j=0}^{l-1}$ by

$$R(\theta) = \sum_{j=0}^{l-1} (h_{k+j}-\bar{h}_k)\cos(j\theta+\varphi) \tag{1}$$

For each θ, the phase shift φ is chosen to maximize the right-hand side of Eq. 1. The inner product $R(\theta)$ is proportional to the correlation between the two sequences. In addition, the inner product $R(\theta)$ can be shown by algebraic manipulation to be equal to the magnitude of the discrete Fourier transform of $\{h_{k+j}-\bar{h}_k\}_{j=0}^{l-1}$.

$$R(\theta)=[(\textstyle\sum_{j=0}^{l-1} (h_{k+j}-\bar{h}_k)\cos\,j\theta)^2 +(\sum_{j=0}^{l-1} (h_{k+j}-\bar{h}_k)\sin\,j\theta)^2]^{1/2} \tag{2}$$

Least Squares. By the method of least squares select a, b, and φ that give the best fit of the functional form

$$a+b\cos(j\theta+\varphi)\quad \text{to}\quad \{h_{k+j}-\bar{h}_k\}_{j=0}^{l-1}$$

Let RS be the residual sum of squares defined by

$$RS=\min_{a,b,\varphi}\,\sum_{j=0}^{l-1} [h_{k+j}-\bar{h}_k-a-b\cos(j\theta+\varphi)]^2$$

and compute the square root of the maximum sum of squares accounted for by

$$S(\theta)=[\textstyle\sum_{j=0}^{l-1} (h_{k+j}-\bar{h}_k)^2 - RS]^{1/2}$$

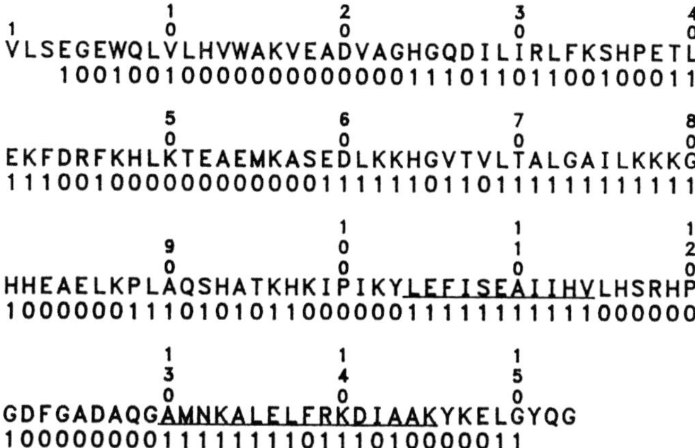

Fig. 1. Sperm whale sequence showing residues centered on blocks of high intensity periodicity. *1* indicates the center of seven residue blocks with dominant periodicity of hydrophobicity in the range of $100\pm20°$, corresponding to that of an α-helix. Numerical values for hydrophobicity were assigned according to the scale of KYTE and DOOLITTLE (1982). Blocks tested around observed T-cell antigenic sites are underlined. Reproduced from DELISI and BERZOFSKY (1985) with permission

For a given interval $\{h_{k+j}-\bar{h}_k\}_{j=0}^{l-1}$ the value θ_R that maximizes $R(\theta)$ and θ_S that maximizes $S(\theta)$ are the frequencies to be associated with the interval. For large values of $l(\geq 20$, say), $\theta_S=\theta_R$, but for the block lengths that we consider (particularly for $l=7$), θ_S and θ_R may differ by as much as $10°$. The value of θ_S may be considered the more accurate of the two in detecting a periodicity in $\{h_{k+j}\}_{j=0}^{l-1}$, for when $\{h_{k+j}\}_{j=0}^{l-1}$ is itself a harmonic sequence $\{\cos(j\theta_0+\varphi)\}_{j=0}^{l-1}$, θ_S will be exactly θ_0 but θ_R may differ from θ_S by as much as $8°$ when θ_S is between $50°$ and $150°$ and as much as $20°$ for other values of θ_0.

Like $R(\theta)$, the function $S(\theta)$ has a similar, although more complex, expression that depends on θ alone. Once programmed into a computer, either method may be used with ease. We speak of the "intensity at θ" as being $R(\theta)$ or $S(\theta)$, depending on the method used, and in the remaining discussion, $I(\theta)$ will denote the intensity at θ, either $R(\theta)$ or $S(\theta)$.

Thus, to identify amphipathic segments with 3.6 residue periodicity, we analyze the complete sequence h_1, h_2, ..., h_n by computing for each overlapping block $\{h_{k+j}\}_{j=0}^{l-1}$ (of fixed length $l=7$ or 11) the angle θ_k at which the intensity $I(\theta)$ is a maximum. When $80°\leq\theta_k\leq120°$, we assign a 1 to the midpoint of the block indicating α-helical periodicity; otherwise we assign a 0 to the midpoint (Fig. 1). An isolated 1 in a sequence of 0s and 1s suggests that whatever the amphipathic α-helix character of the interval that generated the 1, it is disrupted by a shift of one residue in either direction, and thus would seem to be insufficient to stabilize an α-helical structure. A string of five or more consecutive 1s, on the other hand, signals some tendency toward stabilizing an amphipathic α-helical structure, and indicates a candidate antigenic site. In some cases, the sequence of hydrophobicity values fluctuates about the mean of those values with the correct periodicity but does not fluctuate between hydrophilic and hydrophobic values; that is, the residues are all hydrophobic or all hydrophilic. Thus the segment cannot form an amphipathic structure and is excluded from consideration.

In the original study, we used the Fourier analysis approach to examine the periodicity of hydrophobicity of overlapping seven-residue blocks covering the sequences of six proteins for which T-cell sites had been reported (DeLisi and Berzofsky 1985). Ten of the 12 T-cell sites fell into or overlapped with regions of the sequence for which the most intense periodicity was close to that of an α-helix ($100 \pm 20°$, corresponding to $360°/3.6$ residues per turn). These included two of two sites for myoglobin (Fig. 1), two of three for lysozyme, one of one for ovalbumin, two of two for insulin, and three of three for influenza hemagglutinin. Another site was compatible with an amphipathic 3_{10} helix (cytochrome c carboxyl terminal), and the last site with sequential amphipathicity in which a hydrophilic sequence is adjacent to a hydrophobic sequence. Even though the sites are not all α-helical in the native proteins, what may be more important is the propensity to form an amphipathic α-helix after the peptide has been cleaved from the protein during processing and is reexpressed on the surface of the presenting cell, an amphipathic environment which might stabilize such a structure. Other amphipathic structures may also be effective T-cell epitopes (Godfrey et al. 1984).

Since the original study (DeLisi and Berzofsky 1985) several additional epitopes recognized by T cells have come to light. These include two sites of beef cytochrome c (Corradin et al. 1983; G.P. Corradin, personal communication), residues 120–132 of the pre-S region of hepatitis B surface antigen (Milich et al. 1986), and residues 140–160 of VP1 of the foot and mouth disease virus (Francis et al. 1985). All of these correspond to candidate antigenic sites predicted on the basis of their periodicity of hydrophobicity near 3.6 residues per cycle as described above. In addition, a third T-cell epitope predicted in the original analysis of sperm whale myoglobin (DeLisi and Berzofsky 1985) in the segment of residues 59–80 has recently been confirmed experimentally (A. Livingstone, J. Rothbard, and C.G. Fathman, personal communication). Thus, as the limited data base expands, the theory appears to be holding up well. The probability of all these known sites correlating with predicted amphipathic segments by chance alone, taking into account the fraction of each sequence containing amphipathic segments, is less than 0.005. This striking correlation may be an important clue to a fundamental property of T-cell recognition and may provide a general approach for predicting T-cell epitopes, such as for the preparation of synthetic vaccines.

The present algorithm, although a functional working model, has not yet been optimized. We are currently trying to refine and extend the algorithm, taking into account what is known of antigen processing and presentation and T-cell recognition, but trying to retain an important strength of the current algorithm, namely that it requires knowledge only of the primary sequence. We have been examining several additional criteria for evaluating potential

T-cell antigenic sites. Some of these are extensions of or improvements in the prediction of amphipathic segments, while others are distinct from amphipathicity, as noted below.

Among the approaches considered to refine the assessment of amphipathicity, we first sought to optimize the hydrophobicity scale which is central to the whole analysis. After evaluation of over 15 published scales, we have principally employed two hydrophobicity scales, that of KYTE and DOOLITTLE (1982) and the "mean fractional area loss" scale of ROSE et al. (1985). We have found the scale of ROSE et al. to be particularly effective in associating high intensity around $\theta = 100°$ with known α-helical structures (CORNETTE et al. to be published).

Second, we have evaluated more explicit use in the algorithm of the intensity near 100°, as defined above, in addition to frequency. The original analysis considered the frequency of highest intensity for a given block, but did not make use of the actual intensity value at that frequency, relative to other blocks in the sequence. For instance, one might require that at least one of the 1s in a run of five or more should have an intensity $I(\theta)$ that is in the 95th percentile of intensities among all the blocks in the protein and all angles θ, $80° \leq \theta \leq 120°$, or a similar criterion for a normalized intensity

$$I(\theta)/(\textstyle\sum_{j=0}^{l} h_{k+j}^2)^{1/2}$$

Third, we have compared block lengths of 7 versus 11 residues per block (corresponding to two or three turns of an α-helix). Fourth, the original analysis somewhat arbitrarily selected a minimum length of five consecutive blocks with 3.6 periodicity to consider an interval amphipathic. We have evaluated other length intervals. Perhaps even better, we have tried combining the length and intensity criteria into a single parameter on the grounds that a longer run of low intensity might provide as much stabilization or mutual reinforcement for an amphipathic α-helix as a shorter run of high intensity blocks. One way to do this is to sum the intensities, or some function of the intensities, in the 80° to 120° range, over the string of consecutive blocks with this periodicity, and establish some threshold value for this sum which empirically seems to provide enough internal stabilization of the amphipathic structure. This is the approach we are employing in our current working model (MARGALIT et al. to be published).

In addition to optimizing the assessment of amphipathic periodicity, we are examining other distinct criteria. For instance, a region which for other structural reasons would be unlikely to fold with the periodicity that would make it amphipathic might be a less likely candidate for a T-cell antigenic site. Thus, other criteria for the tendency to form α-helical or β-structures are being employed (KLEIN and DELISI 1986; CHOU and FASMAN 1978; GARNIER et al. 1978). We are also trying to incorporate information about charge, glycosylation sites, and possible enzyme cleavage sites.

The optimum way to evaluate these approaches is by discriminant analysis, essentially minimizing the number of false negative and false positive predictions and maximizing the correct positive and negative predictions. Ideally, one uses one part of the data base as a training set to optimize the criteria, and a second part as a test set to evaluate the success of the method. For T-cell antigenic sites, however, this data base is limited in both size and type. First, the number of known sites is rather small to divide into two subsets and still obtain good statistical evaluations. Second, the data base consists only of regions known to be antigenic sites, but none which have been proven *not* to be sites. Therefore, false positive predictions cannot be detected without enlarging the data base to include nonsites. Moreover, given that a site that is immunodominant in one strain of mice is not recognized at all by another strain, depending on *H-2* type, one would have to examine all strains before a peptide could be declared a nonsite.

Despite these obstacles, this approach promises to yield an algorithm capable of predicting with reasonable accuracy likely sites of proteins recognized by T cells. However, it should be pointed out that not all types of T-cell antigenic sites are necessarily amphipathic. The rules being developed may apply only to a certain class of proteins. Even for those sites which are amphipathic, not all are α-helices. For instance, some sites with sequential amphipathicity consisting of a very hydrophobic sequence adjacent to a very hydrophilic sequence have been reported (ALLEN et al. 1984) and were detected in our earlier analysis (DeLisi and Berzofsky 1985). Nevertheless, a surprisingly high percentage of sites so far reported are amphipathic with periodicity near that of an α-helix. Therefore, the method may have widespread usefulness, both in pointing toward some fundamental aspect of the chemistry of helper T-cell recognition of antigen in association with class II MHC molecules and in aiding the choice of antigen sites for applied purposes such as production of synthetic vaccines.

As we attempt to move from analysis to prediction, it should be emphasized that while T-cell antigenic sites may tend to be amphipathic, the converse, that most amphipathic structures are preferred T-cell sites, remains to be examined. Certainly, not all amphipathic segments of proteins are immunodominant sites in any single strain or individual. As we have already seen, the class II MHC antigens of an animal (or person) determine which of the potential antigenic sites is actually immunodominant in that individual. For mammalian proteins, self-tolerance may further limit the available repertoire of T cells. Even if amphipathicity is an important property of many T-cell antigenic sites, it is not sufficient by itself to define a site.

In conclusion, the joint requirements for recognition of antigen on the surface of another cell in association with an Ia histocompatibility molecule and the proteolytic processing or unfolding of native globular proteins which appears to be necessary for this interaction may explain why the T-cell repertoire appears to be highly skewed compared to the antibody repertoire and why T cells do not seem to be specific for native conformation as antibodies frequently are. For synthetic vaccines to immunize for a memory or anamnestic response to the whole protein or pathogen, it is necessary to immunize helper T cells as well as B cells. Immunization with a peptide seen only by B cells, attached to an unrelated carrier protein eliciting carrier-specific T-cell help, may be no better than passive transfer of antibody. Therefore, one should take into consideration these requirements for T-cell recognition in the design of synthetic vaccines. Analysis for amphipathic periodicity may be a first step in this endeavor.

References

Allen PM, Strydom DJ, Unanue ER (1984) Processing of lysozyme by macrophages; identification of the determinant recognized by two T cell hybridomas. Proc Natl Acad Sci USA 81:2489–2493

Arnon R (1984) Antigenic determinants of proteins and synthetic vaccines. In: Schechter AN, Dean A, Goldberger RF (eds) Impact of protein chemistry on the biomedical sciences. Academic, Orlando, p 187

Benacerraf B (1978) A hypothesis to relate the specificity of T lymphocytes and the activity of I region-specific *Ir* genes in macrophages and B lymphocytes. J Immunol 120:1809–1812

Benjamin DC, Berzofsky JA, East IJ, Gurd FRN, Hannum C, Leach SJ, Margoliash E, Michael JG, Miller A, Prager EM, Reichlin M, Sercarz EE, Smith-Gill SJ, Todd PE, Wilson AC (1984) The antigenic structure of proteins: a reappraisal. Annu Rev Immunol 2:67–101

Berkower I, Buckenmeyer GK, Gurd FRN, Berzofsky JA (1982) A possible immunodominant epitope recognized by murine T lymphocytes immune to different myoglobins. Proc Natl Acad Sci USA 79:4723–4727

Berkower I, Matis LA, Buckenmeyer GK, Gurd FRN, Longo DL, Berzofsky JA (1984) Identification of distinct predominant epitopes recognized by myoglobin specific T cells under the control of different *Ir* genes and characterization of representative T cell clones. J Immunol 132:1370–1378

Berkower I, Kawamura H, Matis LA, Berzofsky JA (1985) T cell clones to two major T cell epitopes of myoglobin: Effect of I-A/I-E restriction on epitope dominance. J Immunol 135:2628–2634

Berkower I, Buckenmeyer GK, Berzofsky JA (1986) Molecular mapping of a histocompatibility-restricted immunodominant T cell epitope with synthetic and natural peptides: implications for T cell antigenic structure. J Immunol 136:2498–2503

Berzofsky JA (1980) Immune response genes in the regulation of mammalian immunity. In: Goldberger RF (ed) Biological regulation and development, vol II. Plenum, New York, pp 467–594

Berzofsky JA (1985a) Intrinsic and extrinsic factors in protein antigenic structure. Science 229:932–940

Berzofsky JA (1985b) The nature and role of antigen processing in T cell activation. In: Cruse J, Lewis R (eds) The year in immunology 1984–85. Karger, Basel, pp 18–24

Berzofsky JA (1986) *Ir* genes: antigen-specific genetic regulation of the immune response. In: Sela M (ed) The antigens. Academic, New York (to be published)

Berzofsky JA, Hicks G, Fedorko J, Minna J (1980) Properties of monoclonal antibodies specific for determinants of a protein antigen, myoglobin. J Biol Chem 255:11188–11191

Berzofsky JA, Buckenmeyer GK, Hicks G, Gurd FRN, Feldmann RJ, Minna J (1982) Topographic antigenic determinants recognized by monoclonal antibodies to sperm whale myoglobin. J Biol Chem 257:3189–3198

Cease KB, Berkower I, York-Jolley J, Berzofsky JA (1986) Characterization of differing T cell specificities for an immunodominant site in sperm whale myoglobin using synthetic peptides. Fed Proc 45:619

Chou PY, Fasman GD (1978) Empirical predictions of protein conformation. Annu Rev Biochem 47:251–276

Cornette JL, Cease KB, Margalit H, Spouge JL, Berzofsky JA, DeLisi C (1986) Hydrophobicity scales and computational techniques for detecting amphipathic structures in proteins. In preparation

Corradin GP, Juillerat MA, Vita C, Engers HD (1983) Fine specificity of a BALB/c T cell clone directed against beef apo cytochrome *c*. Mol Immunol 20:763–768

DeLisi C, Berzofsky JA (1985) T cell antigenic sites tend to be amphipathic structures. Proc Natl Acad Sci USA 82:7048–7052

DeLisi C, Cornette J, Margalit H, Cease K, Spouge J, Berzofsky JA (1986) The role of amphipathicity as an indicator of T cell antigenic sites on proteins. In: Sercarz EE, Berzofsky JA (eds) Immunogenicity of protein antigens: repertoire and regulation. CRC, Boca Raton (to be published)

Dorow DS, Shi P-T, Carbone FR, Minasian E, Todd PEE, Leach SJ (1985) Two large immunogenic and antigenic myoglobin peptides and the effects of cyclisation. Mol Immunol 22:1255–1264

East IJ, Hurrell JGR, Todd PEE, Leach SJ (1982) Antigenic specificity of monoclonal antibodies to human myoglobin. J Biol Chem 257:3199–3202

Francis MJ, Fry CM, Rowlands DJ, Brown F, Bittle JL, Houghten RA, Lerner RA (1985) Immunologic priming with synthetic peptides of foot and mouth disease virus. J Gen Virol 66:2347–2354

Garnier J, Osguthorpe DJ, Robson B (1978) Analysis of the accuracy and implications of simple methods for predicting the secondary structure of globular proteins. J Mol Biol 120:97–120

Gell PGH, Benacerraf B (1959) Studies on hypersensitivity. II. Delayed hypersensitivity to denatured proteins in guinea pigs. Immunology 2:64–70

Godfrey WL, Lewis GK, Goodman JW (1984) The anatomy of an antigen molecule: Functional subregions of L-tyrosine-p-azobenzenearsonate. Mol Immunol 21:969–978

Kimoto M, Fathman CG (1980) Antigen-reactive T cell clones. I. Transcomplementing hybrid I-A-region gene products function effectively in antigen presentation. J Exp Med 152:759–770

Klein P, DeLisi C (1986) Prediction of protein structural class from the amino acid sequence. Biopolymers (to be published)

Kyte J, Doolittle RF (1982) A simple method for displaying the hydropathic character of a protein. J Mol Biol 157:105–132

Lando G, Reichlin M (1982) Antigenic structure of sperm whale myoglobin. II. Characterization of antibodies preferentially reactive with peptides arising in response to immunization with the native protein. J Immunol 129:212–216

Lando G, Berzofsky JA, Reichlin M (1982) Antigenic structure of sperm whale myoglobin. I. Partition of specificities between antibodies reactive with peptides and native protein. J Immunol 129:206–211

Lerner RA (1984) Antibodies of predetermined specificity in biology and medicine. Adv Immunol 36:1–44

Margalit H, Spouge JL, Cornette JL, Cease K, DeLisi C, Berzofsky JA (1986) Prediction of T-cell antigenic sites from the primary sequence. (in preparation)

Milich DR, McLachlan A, McNamara MK, Chisari FV, Thornton GB (1986) T and B cell recognition of native and synthetic pre-S region determinants on HBsAg. In: Brown F, Chanock R, Lerner RA (eds) New approaches to immunization. Cold Spring Harbor Laboratories, New York, pp 377–382

Pincus MR, Gerewitz F, Schwartz RH, Scheraga HA (1983) Correlation between the conformation of cytochrome c peptides and their stimulatory activity in a T-lymphocyte proliferation assay. Proc Natl Acad Sci USA 80:3297–3300

Rose GD, Geselowitz AR, Lesser GJ, Lee RH, Zehfus MH (1985) Hydrophobicity of amino acid residues in globular proteins. Science 229:834–838

Rosenthal AS (1978) Determinant selection and macrophage function in genetic control of the immune response. Immunol Rev 40:136–156

Schwartz RH, Fox BS, Fraga E, Chen C, Singh B (1985) The T lymphocyte response to cytochrome c. V. Determination of the minimal peptide size required for stimulation of T cell clones and assessment of the contribution of each residue beyond this size to antigenic potency. J Immunol 135:2598–2608

Staudt LM, Gerhard W (1983) Generation of antibody diversity in the immune response of BALB/c mice to influenza virus hemagglutinin. I. Significant variation in repertoire expression between individual mice. J Exp Med 157:687–704

Streicher HZ, Berkower IJ, Busch M, Gurd FRN, Berzofsky JA (1984) Antigen conformation determines processing requirements for T cell activation. Proc Natl Acad Sci USA 81:6831–6835

Tainer JA, Getzoff ED, Alexander H, Houghten RA, Olson AJ, Lerner RA, Hendrickson WA (1984) The reactivity of anti-peptide antibodies is a function of the atomic mobility of sites in a protein. Nature 312:127–133

Tainer JA, Getzoff ED, Paterson Y, Olson AJ, Lerner RA (1985) The atomic mobility component of protein antigenicity. Annu Rev Immunol 3:501–535

Todd PEE, East IJ, Leach SJ (1982) The immunogenicity and antigenicity of proteins. Trends Biochem Sci 7:212–216

Unanue ER (1984) Antigen-presenting function of the macrophage. Annu Rev Immunol 2:395–428

Watts TH, Gariépy J, Schoolnik GK, McConnell HM (1985) T-cell activation by peptide antigen: effect of peptide sequence and method of antigen presentation. Proc Natl Acad Sci USA 82:5480–5484

Westhof E, Altschuh D, Moras D, Bloomer AC, Mondragon A, Klug A, Van Regenmortel MHV (1984) Correlation between segmental mobility and the location of antigenic determinants in proteins. Nature 311:123–126

Wiley DC, Wilson IA, Skehel JJ (1981) Structural identification of the antibody-binding sites of Hong Kong influenza haemagglutinin and their involvement in antigenic variation. Nature 289:373–378

Peptides as Probes to Study Molecular Mimicry and Virus-Induced Autoimmunity

T. Dyrberg and M.B.A. Oldstone

1   Introduction and Definition

One hypothesis to explain virus (or other microbe)-induced autoimmunity is that an immunogenic determinant of a virus (considered as foreign material) can incite the formation of antibodies or effector lymphocytes, which in turn may react with homologous epitopes on a host protein. One variation on this theme is molecular mimicry, which we define as the presence of epitopes, either linear or conformational, shared between virus and host proteins. When the shared determinants are identical, the organism will be tolerant to the viral epitope by recognizing it as self. Homologous but nonidentical determinants differing in one or more amino acids could be foreign enough to elicit an immune response. The immune response initiated against the foreign material may, however, react not only with it but also with the closely homologous "self" host protein. Depending on the biologic function of the host protein which is recognized and the magnitude of the immune response elicited, the outcome of the interaction may be disease.

2   Mechanism

What determines the outcome of an immune response against a viral epitope mimicking an epitope on a host protein is not well understood, but several factors are believed to be of importance. It may be presumed on the basis of the concept of self tolerance that no cross-reacting antibodies or lymphocytes

Department of Immunology, Research Institute of Scripps Clinic, La Jolla, CA 92037, USA

Current Topics in Microbiology and Immunology, Vol. 130
© Springer-Verlag Berlin · Heidelberg 1986

are induced if an immunogenic determinant on a virus is identical to that on a host protein. An exception might be a state of polyclonal B lymphocyte activation or viral-induced destruction of regulatory T suppressor cells. These events are not discussed in this chapter but are presented elsewhere for the interested reader (NOTKINS and OLDSTONE 1986). Whether an immune response against a viral immunogenic determinant that is partly homologous with a host protein results in cross-reacting antibodies appears to depend on the extent of homology between the two epitopes and possibly on the immune response genes of the major histocompatibility complex (MHC). Thus, in man, low and high responders to several microbial antigens have been identified and correlated with the presence of specific MHC class II antigens (SASAZUKI et al. 1980; LEHNER et al. 1981; VAN EDEN et al. 1983; NOSE et al. 1981). Furthermore, antibodies formed against a viral epitope may, by cross-reacting with a host protein, lead to immune recognition of that self antigen. This event is dependent on a break in the tolerance of the immune system, either at the B or at the T-lymphocyte level (WEIGLE 1980); thereafter, the immune response against the self antigen can continue without the presence of the initiating viral antigen, provided that the self antigen is available. An infectious agent initiating an autoimmune disorder by such a mechanism may at an early stage be cleared from the host and thus absent both at the clinical onset and during the continuum of disease, i.e., a hit-and-run event. Pragmatically, microbiological analysis utilizing immunochemical reagents or tissue culture techniques, as well as nucleic acid probes would therefore fail to reveal the initiating infectious agent. Hence, it may be important to define biochemically those epitopes with which autoreactive antibodies and/or lymphocytes react during autoimmune disease, and to try to identify homologous microbial epitopes in order to shed light on the etiology of the autoimmune process.

The development of autoimmune responses, most often determined by the presence of autoantibodies, is a relatively common event. Examples that could be cited are the high incidence of autoantibodies in elderly humans (LANGE 1978) and relatives of persons with putative autoimmune insulin-dependent diabetes (DOBERSEN et al. 1980), generally not associated with clinical disease, as well as the presence of natural antibodies to normal cell constituents in normal human sera (GUILBERT et al. 1982). Nevertheless, unless such autoantibodies are directed against biologically important domains of host cell proteins, they must certainly play a limited role, if any, in the pathogenesis of autoimmune disease. The exception is the formation of immune complexes, which can produce immune disease by being trapped in renal glomeruli, arteries, and choroid plexus (OLDSTONE 1984).

The nature of the host protein is of major importance in determining whether a cross-reacting immune response induced by molecular mimicry results in disease. Autoimmune responses directed to a biologically important epitope on a host protein are more likely to cause harm than responses to a nonessential epitope on a protein. The accessibility of the self antigen to the immune reactants is also critical, in that only reactions against circulating or self surface components are likely to be of pathogenic importance. The plasma membrane is impermeable to macromolecules; thus, cytoplasmic antigens in the living cell are not accessible to antibodies or lymphocytes. Only when the cell is lysed will cytoplasmic components be released into body fluids. Finally, the MHC class II specificities may be important not only in determining the magnitude of an immune reponse but also in development of autoimmune disease. Several diseases of proven or suspected autoimmune origin have been shown to be closely associated with specific MHC class II genotypes (SVEJGAARD et al. 1983). Furthermore, viruses are known to regulate the expression of MHC gene products by release of lymphokines, notably γ-interferon (WALLACH et al. 1982).

3 Strategy for Selecting Synthetic Peptides as Probes to Study Molecular Mimicry

The existence of molecular mimicry was initially noted through observations that antibodies raised against microorganisms also recognized unrelated self host proteins. Many examples of antigenic determinants shared between viruses, bacteria or parasites, and host cell proteins are now available (Table 1). For example, in a collaborative study performed at the National Institutes of Health by Abner Notkins and his colleagues, 3%–4% of over 600 monoclonal antibodies directed to various DNA and RNA viruses were shown to recognize antigens on uninfected tissues from a variety of organs (SRINIVASAPPA et al. 1985). We have used synthetic peptides as a probe for studying the role of molecular mimicry in the pathogenesis of diseases of unknown or viral etiology. The background that makes this approach feasible is the rapid advance in the molecular cloning of viral and cellular genes, subsequent determination of amino acid sequences of the encoded proteins, the development of protein sequence data banks, computerized search systems for homologies, and advances in peptide chemistry. Implicit in this system – and a major limiting factor – is that only proteins cloned, sequenced, and present in a data bank are available for comparison.

Analysis of the importance of homologous amino acid sequences in viral and host proteins is best pursued by inducing an immune response to a synthetic peptide comprising the shared sequence. In that way, the immune system being tested responds to a single predetermined antigenic sequence rather than the far more complex structure of a whole self host protein or viral protein (LERNER 1984). Theoretical considerations are used to evaluate whether a given homology between two sequences will lead to the induction of cross-reacting antibodies. Recent studies of the immune response to synthetic peptides have shown that sequences as short as six amino acids can be immunogenic (WILSON et al. 1984). This indicates that the minimal overall length of a shared determinant must be at least six amino acids to result in a cross-reacting immune response. The homology between two amino acid sequences can be assessed by the number of identical residues, relatedness of nonidentical amino acids with respect to hydrophilicity-phobicity and charge, and the calculated hydrophilicity and beta turn potential of the entire sequence. Nevertheless, the finding that antibodies to synthetic peptides differing in only one of 19 amino acids may or may not

Table 1. Examples of molecular mimicry between microorganisms and cellular proteins

Microorganism, protein	Homologous cell or protein	Associated disease	References
Simian virus 40, large T antigen	68 K protein in mammalian cell nucleus		LANE and HOEFFLER 1980
Human T-cell leukemia virus, *env*	Extracellular portion of an HLA-B histocompatibility antigen		CLARKE et al. 1983
Human T-cell lymphotropic virus, type 1, p 19 internal core protein	Normal cells and tissues		PALKER et al. 1985
Murine mammary tumor virus, gp 52	B-lymphocyte sub-population		TAX et al. 1983
Polyoma virus, middle-T antigen	130 K and 30 K cellular protein		ITO et al. 1983
Fujinami sarcoma virus transforming protein	98 K bone marrow cell protein		MATHEY-PREVOT et al. 1982
Simian sarcoma virus, p28^{sis}	Human platelet-derived growth factor		ROBBINS et al. 1983
Vaccinia virus, hemagglutinin	Cytoskeleton proteins	Immune complex formation and deposition	DALES et al. 1983
Vaccinia virus, 19 K protein	Epidermal growth factor		BLOMQUIST et al. 1984; REISNER 1985
Coxsackie virus B-3	Myocyte antigens	Virus-induced myocarditis	HUBER and LODGE 1984; WOLFGRAM et al. 1985
Theiler's virus, neutralizing epitope	Galactocerebroside	Demyelination	FUJINAMI and POWELL (1986)
Epstein-Barr virus, nuclear antigen	62 K cellular protein		LUKA et al. 1984
Adenovirus, E1b protein	A-gliadin (in wheat gluten)	Celiac disease	KAGNOFF et al. 1984
Adenovirus, E-3 glycoprotein	β-chain of HLA-DR		CHATTERJEE and MAIZEL 1984
Measles virus phosphoprotein, herpes simplex virus type 1 protein	Human cell intermediate filament protein	Autoantibody appearance during virus infection	FUJINAMI et al. 1983
Measles virus F protein	Cellular stress proteins		SHESHBERADARAN and NORRBY 1984
Simian virus, HN glycoprotein	Purkinje cellular cytoplasmic antigen		GOSWAMI et al. 1985
Hepatitis B virus, polymerase	Rabbit myelin basic protein	Experimental allergic encephalomyelitis	FUJINAMI and OLDSTONE 1985

Table 1 (continued)

Microorganism, protein	Homologous cell or protein	Associated disease	References
Streptococci, M protein	Cardiac myosin	Rheumatic fever	DALE and BEACHEY 1985; KRISHER and CUNNINGHAM 1985
Menningococcus group B, capsule proteins	Human and rat brain glycopeptides	Meningitis	FINNE et al. 1983
Escherichia coli, Proteus vulgaris, Klebsiella pneumoniae	α-subunit of acetylcholine receptor	Myasthenia gravis	STEFANSSON et al. 1985
Trypanosoma cruzi	Mammalian neurones and cardiac muscle	Chagas' disease	WOOD et al. 1982

bind to the whole protein (ALEXANDER et al. 1983) indicates that a cross-reacting immune response may not necessarily be generated, in spite of extensive homology between two sequences.

Biologically important antigens, or their epitopes, are known in several naturally occurring or experimentally induced immune diseases. For example, in allergic encephalomyelitis (EAE) the encephalitogenic sites of myelin basic protein is known for several animal species. Utilizing a computer search, FUJINAMI and OLDSTONE (1985) found that hepatitis B virus polymerase shared six consecutive amino acids with the ten amino acid encephalitogenic site of rabbit myelin basic protein (MBP). After rabbits were immunized with the ten amino acid sequence from the hepatitis B virus polymerase that contained the six shared amino acids, their sera and peripheral blood mononuclear cells were observed to react with intact MBP. Hence, a peptide (ten amino acids, in other experiments eight amino acids) whose sequence was derived from a virus protein elicited a humoral and cellular immune response that reacted not only with the inducing viral peptide and the rabbit encephalitogenic peptide, but also with the whole native MBP. Most importantly, central nervous system tissue taken from rabbits immunized by the peripheral route histologically resembled that found in EAE. Perivascular round cell infiltrates were restricted to the central nervous system and found primarily in the subependymal area, under the third ventricle, and in scattered sites in the midbrain and forebrain. This study represented the first direct evidence to support the hypothesis that molecular mimicry can cause autoimmune disease. Furthermore, it indicated a role of virus infection in the triggering and production of antibodies and lymphocytes with the potential of cross-reacting with a biologically important self protein in vivo.

At present the study of the cross-reactivity of hepatitis B viral peptide and the encephalitogenic site of MBP in the rabbit is the only evidence for autoimmune disease induction by molecular mimicry; however, there are many suggestions that similar events may occur in other diseases (see Table 1). For example, a region of amino acid sequence homology has been demonstrated between A-gliadin, a wheat gluten component which is known to activate celiac disease, and the 54 K E1b protein of human adenovirus, which can usually be isolated from the intestinal tract (KAGNOFF et al. 1984). Antisera reactive with the E1b virus protein cross-reacted with the homologous synthetic A-gliadin peptide, suggesting that an immune response against this viral antigenic determinant may be involved in the pathogenesis of celiac disease (KAGNOFF et al. 1984).

4   Using Molecular Mimicry to Understand Human Diseases Like Myasthenia Gravis and Diabetes Mellitus: An Experimental Approach Utilizing Synthetic Peptide Probes

Myasthenia gravis is a neurologic disease caused by a critical reduction in the number of nicotinic acetylcholine receptors. The pathogenesis of the disease is believed to be a humoral autoimmune reaction against the receptor (LINDSTROM 1985). The etiology is unknown, but virus infections have previously been implicated as initiating factors in the disease (DATTA and SCHWARTZ 1974; KORN and ABRAMSKY 1981). Autoantibodies to the acetylcholine receptor are often found in patients with myasthenia gravis and are primarily directed against the α-chain (LINDSTROM 1985). To test the hypothesis that viruses may be etiologic factors in myasthenia gravis, we employed computer analysis of a protein sequence data base (Protein Identification Resource, Washington D.C.) to search for homologous amino acid sequences between the acetylcholine receptor and virus proteins. In particular, we sought identity with an eight amino acid sequence of the α-chain of the human acetylcholine receptor, proposed to be the major immunogenic region of the receptor (NODA et al. 1983). Our computer search identified several viral proteins showing homology with the test sequence in question. Some of the sequences selected for study are listed in Table 2. The peptides were made on an automated peptide synthesizer, analyzed by high-pressure liquid chromatography, coupled to keyhole limpet hemocyanin

Table 2. Virus protein amino acid sequences showing homology with the α-chain of the human acetylcholine receptor

Protein	Amino acid positions	Amino acid sequence
Human acetylcholine receptor, α-chain	160–167	P E S D Q P D L
Polyoma virus, middle T-antigen	317–324	P E S D Q D Q L
Herpes simplex virus, glycoprotein D	286–293	P N A T Q P E L
	381–388	R E D D Q P S S
Parvovirus H1, VP2 protein	135–142	T E T N Q P D T

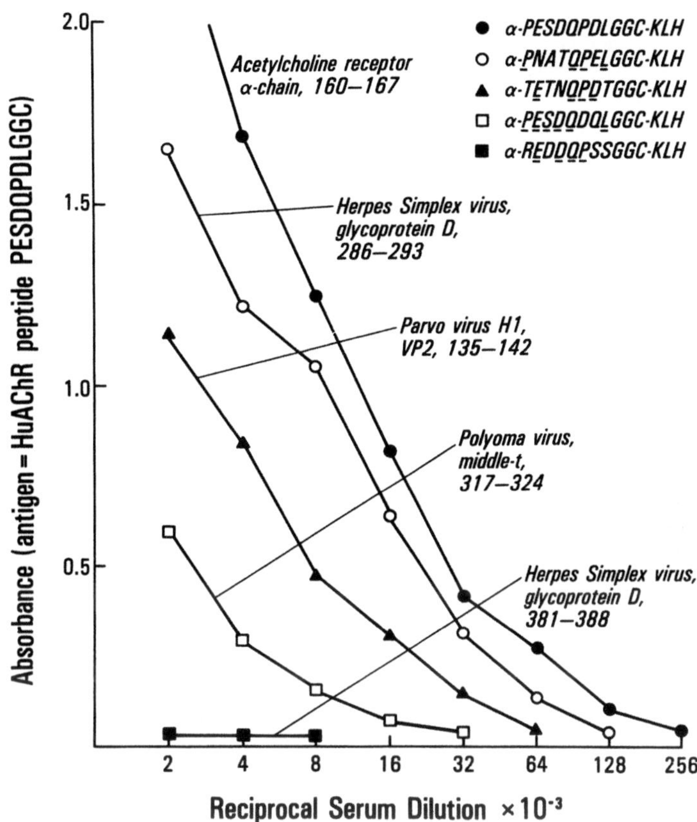

Fig. 1. Rabbit antisera were raised against synthetic peptides; sequences were derived from the α-chain of the human acetylcholine receptor or virus proteins (see Table 2). Antisera cross-reactivities with the acetylcholine receptor peptide were tested in an enzyme-linked immunosorbent assay, using the receptor peptide as antigen

through a COOH-terminal cysteine residue (LIU et al. 1979), and used to immunize rabbits. Antibody production was analyzed by an enzyme-linked immunosorbent assay (ELISA), using uncoupled peptides as antigens. High-titered antisera against the immunizing peptides were produced in all cases. Analysis of the binding of antibodies induced to the viral peptides for the acetylcholine receptor peptide demonstrated a varying degree of cross-reactivity (Fig. 1). Thus, antibody to herpes simplex virus (HSV) glycoprotein D peptide (residues 286–293) was highly cross-reactive, while in contrast HSV peptide (residues 381–388) from the same protein and showing an equivalent degree of homology did not react with the receptor peptide. The polyoma middle-T antigen peptide (residues 317–324) showed the highest degree of homology with the acetylcholine receptor α-chain peptide, but antibody to the polyoma peptide showed a low binding to the acetylcholine receptor peptide (Fig. 1). In a similar vein, RICHMAN and his colleagues (STEFANSSON et al. 1985) showed that several monoclonal antibodies to the α-subunit of the acetylcholine receptor recognized determinants

Table 3. Examples of molecular mimicry between the extramembranal parts of the human insulin receptor and virus proteins

Protein	Amino acid position	Amino acid sequence
Human insulin receptor, α-chain	65–75	R V Y G L E S L K D L
Papilloma virus, E2 protein	75–85	M V L H L E S L K D S
Human insulin receptor, α-chain	586–600	Y V Q T D A T N P S V P L D P
Poliovirus, VP1 protein	624–638	A V E T G A T N P L V P S D T
Human insulin receptor, β-chain	735–743	V P T V A A F P N
Epstein-Barr virus, BV RF1 protein	446–454	R P T V A A D P Q

in proteins from *E. coli, P. vulgaris,* and *K. pneumoniae.* These observations, in concert with ours, indicate that synthetic peptides might be used as an immunoabsorbent to purify unique antibodies from the serum of myasthenic patients. Utilizing this approach, we have begun to find such antibodies, and they are currently being evaluated for their ability to passively transfer myasthenia gravis. By such studies, potential etiologic agents may be uncovered.

Several diseases of unknown etiology, like diabetes mellitus, are associated with the production of autoantibodies. In some cases diabetes is caused by autoantibodies to the insulin receptor (KAHN and HARRISON 1981). The insulin receptor of humans has recently been cloned and sequenced. It contains an extramembranal α-chain and transmembranal β-chain that come from a common precursor (ULLRICH et al. 1985). We have analyzed the α- and β-chains, in overlapping ten amino acid sequences, for homologies to viral proteins. In addition to the reported homologies of the receptor β-chain to retroviral oncogene product with tyrosine kinase activity (ULLRICH et al. 1985), several homologous sequences were apparent between the external part of the α- or β-chains and various viral proteins. Several of these homologies are listed in Table 3. The strategy detailed above for myasthenia gravis is available for analysis of this disease as well as others.

5 Peptide Structure and Immune Responses

The use of peptides for biomedical studies requires analysis of their use as immunogens. We tested the immunogenic effects of peptides when either the NH_2- or COOH-terminal regions were coupled with a carrier protein. For these studies, peptides were synthesized with amino acid sequences identical to those of the acetylcholine receptor and polyoma virus peptide (Table 2). Two glycine residues were added as spacers between the core peptide sequence and a cysteine residue (Table 4). When rabbits were immunized with peptides whose amino terminus was coupled to carrier protein, the resultant antisera titers were comparable to antibodies raised against the homologous peptides coupled at the

Table 4. Homologous peptide sequences from human acetylcholine receptor, α-chain, and polyoma virus, middle-T antigen synthesized with either a NH_2- or COOH-terminal cysteine residue

			160						167			
Human acetylcholine receptor, α-chain	C	G	<u>G</u>	P	E	S	D	Q	P	<u>D</u>	L	
			P	E	S	D	Q	P	D	L	G	G	C

			317						324			
Polyoma virus, middle-T antigen	C	G	G	<u>P</u>	<u>E</u>	<u>S</u>	<u>D</u>	Q	D	Q	<u>L</u>	
				<u>P</u>	<u>E</u>	<u>S</u>	<u>D</u>	Q	D	Q	<u>L</u>	G	G	C

Rabbits were immunized with the peptides after the latter had been coupled to keyhole limpet hemocyanin through the cysteine residue, and the resulting peptide antisera were tested for binding reactivity to the immunizing and homologous peptides

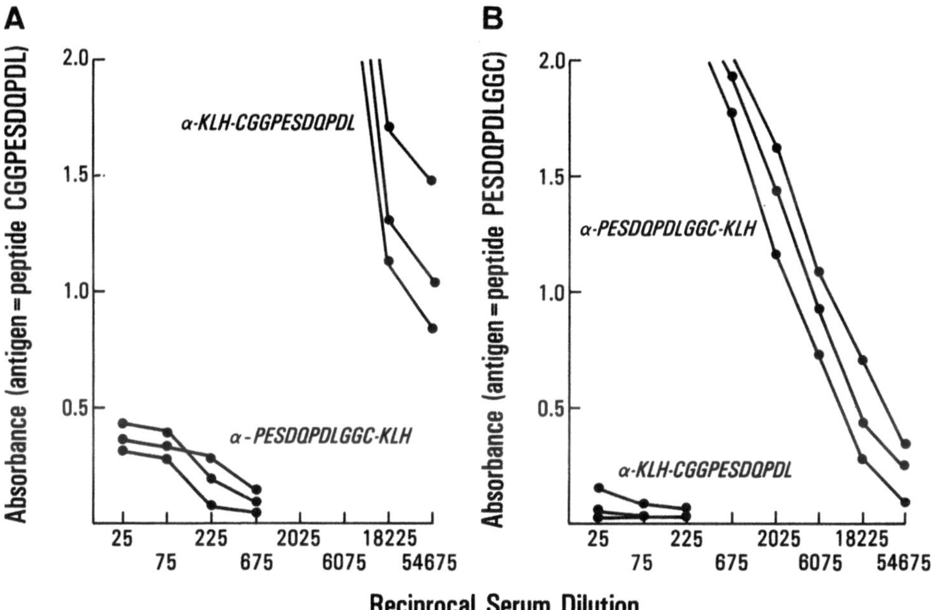

Fig. 2 A, B. Rabbits were immunized with a synthetic peptide; the sequence being derived from the α-chain of the human acetylcholine receptor. Peptides were elongated either NH_2- or COOH-terminal with two glycine residues and one cysteine residue through which the peptide was coupled to keyhole limpet hemocyanin before being used for immunization. Binding of the peptide antisera to either peptide was analyzed in an enzyme-linked immunosorbent assay, using uncoupled peptide as antigens. Binding values have been corrected for nonspecific binding induced by preimmune sera. **A** Antigen, peptide with NH_2 terminal cysteine. **B** Antigen, peptide with COOH terminal cysteine

COOH end. Surprisingly, despite the close homology between the peptides, a remarkable lack of cross-reactivity was observed between the NH_2- and COOH-induced antisera, even at a wide range of serum dilutions (Fig. 2). These results were consistent with all sera analyzed from rabbits immunized with either acetylcholine receptor or polyoma virus peptides. Quantitative absorption studies confirmed these results. Other studies using amino acid spacers in different parts of the peptide showed that charge and polarity are critical for the immunizing peptides (DYRBERG and OLDSTONE 1986). In other instances the coupling of the peptide to the carrier protein may interfere with the binding pocket of the antibody. Hence, the coupling of a peptide to carrier protein, usually required to assure appropriate immunization, might affect the charge or polarity of an "immunogenic" part of a peptide.

6 Concluding Remarks

Observations and analysis of homologies between a variety of important host proteins like the insulin receptor, the acetylcholine receptor, the encephalitogenic sites of myelin basic protein and myelin proteolipid on one side and infectious agents on the other offer a unique way of probing for possible etiologic agents of these and other diseases.

Virus infection is often associated with the development of autoimmunity. This can occur by a variety of mechanisms such as polyclonal B-cell activation, production of autoantibodies via anti-idiotype feedback devices, infection of selected immunocyte subpopulations, or interferon-altered expression of MHC gene products. Molecular mimicry between viral proteins and host cell proteins is an additional strategy by which viruses can induce autoimmunity. The observations that 3%–4% percent of monoclonal antibodies directed against a variety of DNA and RNA viruses react against normal "self" protein indicate that molecular mimicry is a common event. In addition to cross-reacting antibodies, T-lymphocyte clones that react both with virus and self proteins have been described, i.e., measles virus and MBP, rubella and MBP, etc. (JOHNSON and GRIFFIN 1986).

The use of synthetic peptides provides a tool to analyze those amino sequences involved in such cross-reactions. However, the demonstration of amino acid sequences shared between viral and host cell proteins is a first step. Thereafter, induction of cross-reacting antibodies and lymphocytes by the shared peptide sequences and analysis of their binding to native protein are needed. Finally, to assign a role for molecular mimicry in the pathogenesis of a particular autoimmune disease, it is necessary to demonstrate that adoptive transfer of immunologic reactants to the self protein can cause pathologic lesions characteristic for that specific autoimmune disease.

The investigate approaches described in this article may provide two other useful pieces of information. First, by uncovering molecular mimicry and generation of cross-reactive antibodies, it should be possible to synthesize peptides to determine the molecular basis of tolerance and the minimal number of amino acids in a peptide needed to break the tolerance. Second, the probing of specific diseases associated with the generation of autoreactive antibodies or T-cell clones by means of the peptide technology described here, together with the utilization of monoclonal antibodies, antibodies to predetermined sequences, and cDNA fragments expressed in λ vectors, may well uncover pathogenetically important sites in human disease. For example, this approach may be used to uncover the encephalitogenic site in human MBP or in human proteolipid protein of myelin.

Acknowledgments. This is publication number 4209-IMM from the Department of Immunology, Scripps Clinic and Research Foundation, La Jolla, California 92037. This research was supported in part by USPHS grants AI-07007, AG-04342, and NS-12428. Thomas Dyrberg is the recipient of a senior research fellowship from the University of Copenhagen, Denmark, and a travel grant from the Fulbright Program of the Council for International Exchange of Scholars.

References

Alexander H, Johnson DA, Rosen J, Jerabek L, Green N, Weissman IL, Lerner RA (1983) Mimicking the alloantigenicity of proteins with chemically synthesized peptides differing in single amino acids. Nature 306:697–699

Blomquist MC, Hunt LT, Barker WC (1984) Vaccinia virus 19-kilodalton protein: relationship to several mammalian proteins, including two growth factors. Proc Natl Acad Sci USA 81:7363–7367

Chatterjee D, Maizel JV, Jr (1984) Homology of adenoviral E3 glycoprotein with HLA-DR heavy chain. Proc Natl Acad Sci USA 81:6039–6043

Clarke MF, Gelmann EP, Reitz MS, Jr (1983) Homology of human T-cell leukaemia virus envelope gene with class I HLA gene. Nature 305:60–62

Dale JB, Beachey EH (1985) Epitopes of streptococcal M proteins shared with cardiac myosin. J Exp Med 162:583–591

Dales S, Fujinami RS, Oldstone MBA (1983) Infection with vaccinia favors the selection of hybridomas synthesizing autoantibodies against intermediate filaments, one of them cross-reacting with the virus hemagglutinin. J Immunol 131:1546–1553

Datta SK, Schwartz RS (1974) Infectious(?) myasthenia. N Engl J Med 291:1304–1305

Dobersen MJ, Scharff JE, Ginsberg-Fellner F, Notkins AL (1980) Cytotoxic autoantibodies to beta cells in the serum of patients with insulin-dependent diabetes mellitus. N Engl J Med 303:1493–1498

Dyrberg T, Oldstone MBA (1986) Peptides as antigens: Importance of orientation. J Exp Med, in press

Finne J, Leinonen M, Makela PH (1983) Antigenic similarities between brain components and bacteria causing meningitis. Lancet ii:355–357

Fujinami RS, Oldstone MBA (1985) Amino acid homology and immune responses between the encephalitogenic site of myelin basic protein and virus: A mechanism for autoimmunity. Science 230:1043–1045

Fujinami RS, Powell H (1986) A monoclonal antibody that neutralizes Theiler's virus, reacts with galactocerebroside and induces demyelination (to be published)

Fujinami RS, Oldstone MBA, Wroblewska Z, Frankel ME, Koprowski H (1983) Molecular mimicry in virus infection: cross-reaction of measles virus phosphoprotein or of herpes simplex protein with human intermediate filaments. Proc Natl Acad Sci USA 80:2346–2350

Goswami K, Morris R, Rastogi S, Lange L, Russell W (1985) A neutralizing monoclonal antibody against a paramyxovirus reacts with a brain antigen. J Neuroimmunol 9:99–108

Guilbert B, Dighiero G, Avrameas S (1982) Naturally occurring antibodies against nine common antigens in human sera. J Immunol 128:2779–2787

Huber SA, Lodge PA (1984) Coxsackie-virus B-3 myocarditis in BALB/c mice: evidence for autoimmunity to myocyte antigens. Am J Path 116:21–29

Ito Y, Hamagishi Y, Segawa K, Dalianis T, Appella E, Willingham M (1983) Antibodies against nonapeptide of polyomavirus middle T antigen: cross-reaction with a cellular protein(s). J Virol 48:709–720

Johnson RT, Griffin ED (1986) Virus-induced autoimmune demyelinating disease of the central nervous system. In: Notkins AL, Oldstone MBA (eds) Concepts in viral pathogenesis, vol 2. Springer, New York Heidelberg Berlin Tokyo

Kagnoff MF, Austin RK, Hubert JJ, Bernardin JE, Kasarda DD (1984) Possible role for a human adenovirus in the pathogenesis of celiac disease. J Exp Med 160:1544–1557

Kahn CR, Harrison LC (1981) Insulin receptor autoantibodies. In: Randle PJ et al. (eds) Carbohydrate metabolism and its disorders, vol 3. Academic, London, pp 279–330

Korn IL, Abramsky O (1981) Myasthenia gravis following viral infection. Eur J Neurol 20:435–439

Krisher K, Cunningham MW (1985) Myosin: a link between streptococci and heart. Science 227:413–415

Lane DP, Hoeffler WK (1980) SV40 large T shares an antigenic determinant with a cellular protein of molecular weight 68,000. Nature 288:167–170

Lange CF (1978) Immunology of aging. Prog Clin Path 7:119–136

Lehner T, Lamb J, Welsh KL, Batchelor RJ (1981) Association between HLA-DR antigens and helper cell activity in the control of dental caries. Nature 292:770–772

Lerner RA (1984) Antibodies of predetermined specificity in biology and medicine. Adv Immunol 36:1–44

Lindstrom J (1985) Immunobiology of myasthenia gravis, experimental autoimmune myasthenia gravis and Lambert-Eaton syndrome. Annu Rev Immunol 3:109–131

Liu F-T, Zinnecker M, Hamaoka T, Katz DH (1979) New procedures for preparation and isolation of conjugates of proteins and a synthetic copolymer of D-amino acids and immunochemical characterization of such conjugates. Biochemistry 18:690–697

Luka J, Kreofsky T, Pearson GR, Hennessy K, Kieff E (1984) Identification and characterization of a cellular protein that cross-reacts with the Epstein-Barr virus nuclear antigen. J Virol 52:833–838

Mathey-Prevot R, Hanafusa H, Kawai S (1982) A cellular protein is immunologically cross-reactive with and functionally homologous to the Fujinami sarcoma virus transforming protein. Cell 28:897–906

Noda M, Furutani Y, Takahashi H, Toyosato M, Tanabe T, Shimizu S, Kikyotani S, Kayano T, Hirose T, Inayama S, Numa S (1983) Cloning and sequence analysis of calf cDNA and human genomic DNA encoding alpha-subunit precursor of muscle acetylcholine receptor. Nature 305:818–823

Nose Y, Komori K, Inoye H, Nomura K, Yamamura M, Tsuji K (1981) Role of macrophages in T lymphocyte response to candida allergen in man with special reference to HLA-D and DR. Clin Exp Immunol 45:152–157

Notkins AL, Oldstone MBA (eds) (1984) Concepts in viral pathogenesis. Springer, New York Heidelberg Berlin Tokyo

Notkins AL, Oldstone MBA (eds) (1986) Concepts in viral pathogenesis II. Springer, New York Heidelberg Berlin Tokyo

Oldstone MBA (1984) Virus-induced immune complex formation and disease: definition, regulation, importance. In: Concepts in viral pathogenesis. Springer, New York Heidelberg Berlin Tokyo, pp 201–209

Palker T, Scearce R, Ho W, Copeland T, Oroszlan S, Popovic M, Haynes B (1985) Monoclonal antibodies reactive with human T cell lymphotropic virus (HTLV) P19 internal core protein: Cross-reactivity with normal tissues and differential reactivity with HTLV types I and II. J Immunol 135:247–254

Reisner AH (1985) Similarity between the vaccinia virus 19K early protein and epidermal growth factor. Nature 313:801–803

Robbins KC, Antoniades HN, Devard SG, Hunkapiller MW, Aaronson SA (1983) Structural and immunological similarities between simian sarcoma virus gene product(s) and human platelet-derived growth factor. Nature 305:605–608

Sasazuki T, Chta N, Kaneoka R, Kojima S (1980) Association between an HLA haplotype and low responsiveness to shistosomal worm antigen in man. J Exp Med 152 (suppl):314s–318s

Sheshberadaran H, Norrby E (1984) Three monoclonal antibodies against measles virus F protein cross-react with cellular stress proteins. J Virol 52:995–999

Srinivasappa J, Saegusa J, Prabhakar BS, Gentry MK, Buchmeier MJ, Wiktor TJ, Koprowski H, Oldstone MBA, Notkins AL (1986) Molecular mimicry: frequency of reactivity of monoclonal antiviral antibodies with normal tissues. J Virol 57:397–401

Stefansson K, Dieperink ME, Richman DP, Gomez CM, Marton LS (1985) Sharing of antigenic determinants between the nicotonic acetylcholine receptor and proteins in *Escherichia coli*, *Proteus vulgaris* and *Klebsiella pneumoniae*. N Engl J Med 312:221–225

Svejgaard A, Platz P, Ryder L (1983) HLA and disease 1982 – a survey. Immunol Rev 70:193–218

Tax A, Ewert D, Manson LA (1983) An antigen cross-reactive with gp52 of mammary tumor virus is expressed on a B cell subpopulation of mice. J Immunol 130:2368–2371

Ullrich A, Bell JR, Chen EY, Herrera R, Petruzelli LM, Dull TJ, Gray A, Coussens L, Liao Y-C, Tsubokawa M, Mason A, Seeburg PH, Grunfeld C, Rosen OM, Ramachandran J (1985) Human insulin receptor and its relationship to the tyrosine kinase family of oncogenes. Nature 313:756–761

Van Eden W, De Vries RRP, Stanford JL, Rook GAW (1983) HLA-DR3 associated genetic control of response to multiple skin tests with new tuberculin. Clin Exp Immunol 52:287–292

Wallach D, Fellous M, Revel M (1982) Preferential effect of γ-interferon on the synthesis of HLA antigens and their mRNAs in human cells. Nature 299:833–836

Weigle WO (1980) Analysis of autoimmunity through experimental models of thyroiditis and allergic encephalomyelitis. Adv Immunol 30:159–273

Wilson I, Niman HL, Houghten RA, Cherenson AR, Connolly ML, Lerner RA (1984) The structure of an antigenic determinant in a protein. Cell 37:767–778

Wolfgram LJ, Beisel KW, Rose NR (1985) Heart-specific autoantibodies following murine coxsackie-virus B3 myocarditis. J Exp Med 161:1112–1121

Wood JN, Hudson L, Jessell TM, Yamamoto M (1982) A monoclonal antibody defining antigenic determinants on subpopulations of mammalian neurons and *Trypanosoma cruzi* parasites. Nature 296:34–38

The Use of Peptides in Studying Mechanisms of Immune Tolerance*

G.M. Gammon and E.E. Sercarz

1 Introduction

The administration of a peptide to experimental animals can have several different immunologic consequences. The peptide may stimulate helper T cells and possibly other lymphocyte subpopulations to respond. Alternatively, it might induce a state of tolerance, so that no response would be detected initially or on subsequent exposure to the peptide even in a form that stimulates a response in naive animals. Before peptides can be used as immunogens, it is important to understand not only what is required to productively engage the immune system, but also to know under what conditions a peptide can induce tolerance. Certainly it would be counterproductive if a peptide intended to stimulate protective immunity instead caused nonresponsiveness. T-lymphocyte tolerance is more likely to be a problem than B-cell tolerance. The B cells which would recognize a small peptide are not the same ones that would recognize the native protein unless the peptide retained the same conformation it had as part of the whole protein. However, T cells normally interact with denatured antigen derived from processing of the native molecule in antigen-presenting cells (Ziegler and Unanue 1981; Germain 1981; Shimonkevitz et al. 1983), and so there is a large overlap between T cells responding to the native protein and those responding to peptide fragments.

In this chapter, we will discuss the mechanism of T-cell tolerance induction to small synthetic peptides and how this might relate to the use of peptides as vaccines. There are several important questions. How may peptides induce tolerance? Is it possible to prepare peptides that are always immunogenic and never tolerogenic, or can any peptide induce tolerance? If the latter were true, what are the special requirements to induce tolerance?

* This research was supported NSF PCM 83 02679 and NIH AI-11183

Department of Microbiology, University of California, Los Angeles, CA 90024, USA

Although there are many examples of experimental T-lymphocyte tolerance, the underlying mechanisms are poorly understood. Two major mechanisms have been proposed. First, tolerance may be directly caused by the inactivation of lymphocytes by contact with antigen (WEIGLE 1980; WATERS et al. 1979). Why interaction with antigen should cause inactivation in some circumstances rather than stimulation is unclear. Lymphocytes may go through an obligatory phase of development in which contact with antigen prevents further maturation or causes cell death. Alternatively, it has been proposed that a second signal in addition to antigen is needed to stimulate T cells, and if antigen is presented in the absence of this second signal, the T cell is inactivated. Thus, in this model, clonal activation is due to modified antigen presentation. The second major hypothesis to explain tolerance induction is that the tolerogen induces antigen-specific suppressor T cells which block subsequent responses to the antigen (LOBLAY et al. 1983). These mechanisms of clonal inactivation and suppression are not mutually exclusive and probably are both important. For example, tolerance to hen eggwhite lysozyme (HEL) induced in neonatal mice appears to be due to clonal inactivation, whereas tolerance to the same protein induced by intravenous injection of HEL in adult mice is predominantly due to suppression (OKI and SERCARZ 1985).

If tolerance were due to clonal inactivation, then any peptide that induces a response should also have the capacity to induce tolerance. However, if it were due to suppression, a peptide would have to contain a suppressor T cell-inducing determinant (SD) to be tolerogenic. This is important because it should be possible to synthesize immunogenic peptides that lack an SD. Detailed analysis of the proliferative T-cell (Tp) response to several protein antigens, e.g., lysozyme (MANCA et al. 1984; BIXLER et al. 1985), myoglobin (BERKOWER et al. 1984), cytochrome c (SOLINGER et al. 1979), and insulin (ROSENWASSER et al. 1979) has demonstrated that the responding cells are restricted to one or a very few small regions on the immunogen. Even in proteins such as HEL which differ from the mouse equivalent by 40% of its amino acid residues and so would be expected to contain many Tp-inducing determinants, one or two determinants seem to be dominant. This appears to be true for suppressor T cells (Ts) also, as in several systems, e.g., lysozyme (ADORINI et al. 1979), insulin (JENSEN and KAPP 1985), and β-galactosidase (KRZYCH et al. 1983), Ts are specific for a limited range of epitopes. Where both the Tp and Ts specificities for the same protein have been elucidated, they have been found to be distinct and nonoverlapping. The specificities of Tp and Ts induced by small peptides have not been extensively investigated, but appear to be separate as in the case of native proteins. PETERSON, WILNER and THOMAS (personal communication) found that both helper (Th) and suppressor T cells could be generated to a 14 amino acid peptide of human fibrinopeptide B, and that each subpopulation recognized a different epitope which could be distinguished using variant peptides. A similar finding has been reported in the lactic dehydrogenase system (SERVIS et al. 1986). Therefore, because the determinants recognized by different T-cell subsets seem to be distinct, it should be feasible to prepare non-suppressogenic peptides, although these may still be tolerogenic by other mechanisms.

2 Tolerance to HEL: Analysis of Tolerance by Responses to Peptides

As stated above, tolerance to HEL may be due to either active suppression or clonal inactivation (OKI and SERCARZ 1985). Tolerance can be induced in adult B10.A mice by a single intravenous injection of 2 mg HEL in saline. Although challenge with whole HEL fails to prime for an in vitro proliferative response, immunization with certain HEL peptide fragments can induce a response in tolerant mice which is fully recalled by the native protein. Two cyanogen bromide cleavage peptides, L2 (amino acids 13–105) and L3 (amino acids 106–129), both stimulate a proliferative response. Furthermore, AP-HEL (amino acids 4–129), which lacks only the first three N-terminal amino acids also induces a response in HEL-tolerant mice. Since such latent responses can be revealed using peptides, tolerance cannot be solely the result of antigen-induced clonal inactivation. Instead, the data suggest that tolerance is due to suppression by Ts which are induced by a determinant involving the N-terminal region of HEL and which act on Tp specific for other regions of the molecule. Figure 1 is a diagrammatic representation of this. This is another example where Ts- and Tp-inducing determinants on a protein are distinct. In addition, it has been shown that peptides which lack this SD can stimulate latent reactive cells.

Tolerance to HEL can be induced in neonatal mice by a single injection of 100 µg HEL given in a saline/Freund's incomplete adjuvant emulsion. Challenge with the peptide fragments L2 and L3 has been used to test this tolerance, but unlike the case with HEL tolerance induced in the adult, neither stimulates a response. The inability to stimulate any latent reactive cells is strong evidence that in neonatally induced tolerance to HEL, all the HEL-reactive T cells have undergone an inactivation process. However, the peptides used were still quite large and it is possible that additional Ts-inducing determinants which are not utilized in adult tolerance induction exist on each peptide. This problem can be overcome by using small enough synthetic peptides which would bear, at most, a single immunogenic determinant. This novel approach was possible

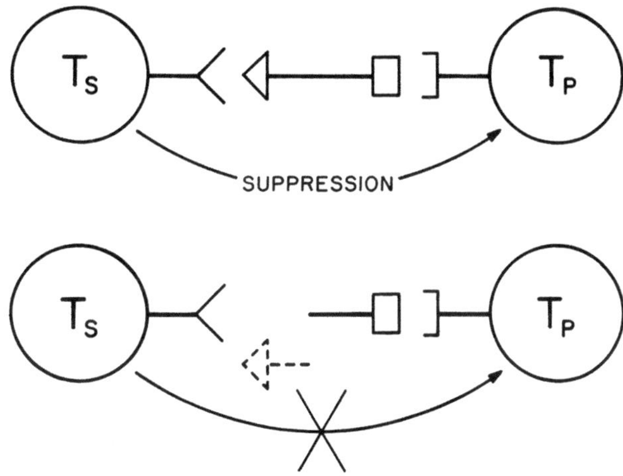

Fig. 1. Removal of a suppressor determinant allows latent responses to develop

through the synthesis of desired peptides in areas of the molecule known to possess proliferation-inducing determinants. Solid phase procedures enables us to prepare a panel of peptides of differing lengths with variants having single amino acid substitutions. Another major advantage of synthetic procedures was the freedom from worry about proteolytic fragments from other regions of the whole protein contaminating peptide preparations, possibly invalidating certain conclusions.

Recently, we have been using small synthetic peptides to induce tolerance in neonatal B10.A mice in order to study the underlying mechanisms (GAMMON et al. 1986). Tolerance can be easily induced by two injections of peptide into the peritoneal cavity at 24 h and 72 h after birth. Each peptide was given in an emulsion with Freund's incomplete adjuvant (FIA) which acts as a reservoir, releasing small amounts of peptide over a long period. Tolerance was assessed at age 8–10 weeks by challenge with the peptide in Freund's complete adjuvant (FCA) and the in vitro proliferative response measured upon culture of lymphocytes from the draining lymph nodes 9 or 10 days later.

3 Tolerance to HEL Peptides: Neonatal Tolerance Induction with Minimal Immunogenic Peptides

T11, a synthetic peptide corresponding to amino acid residues 74–96 of HEL, stimulates a strong in vitro proliferative response in adult B10.A mice. The fine specificity of the T-cell response has been extensively investigated using a panel of synthetic peptides (Table 1). Three long-term T-cell lines were established from T11-immunized mice and individual clones isolated from each. All the clones responded to T11 and either the T11 amino-terminal peptide 74–86 or the carboxy-terminal peptide 85–96 (SHASTRI et al. 1985; N. SHASTRI et al. to be published). Therefore, T11 bears at least two distinct determinants, and lymph node cells from adult B10.A mice immunized with T11 respond in vitro to both peptides 74–86 and 85–96. This experimental system has been used to investigate neonatal tolerance. The first question was whether tolerance could be induced using small synthetic peptides. To test this, neonatal B10.A mice were injected with T11 in a saline/FIA emulsion and challenged at 8 weeks

Table 1. Synthetic HEL peptides used to investigate tolerance. Numbers refer to the amino acid residues of HEL sequence

Synthetic peptide		75					80					85					90					95	
T11(74–96)	N	L	C	N	I	P	C	S	A	L	L	S	S	D	I	T	A	S	V	N	C	A	K
85–96												–	–	–	–	–	–	–	–	–	–	–	–
74–86	–	–	–	–	–	–	–	–	–	–	–	–	–										
74–82	–	–	–	–	–	–	–	–	–														
77–86				–	–	–	–	–	–	–	–	–	–										

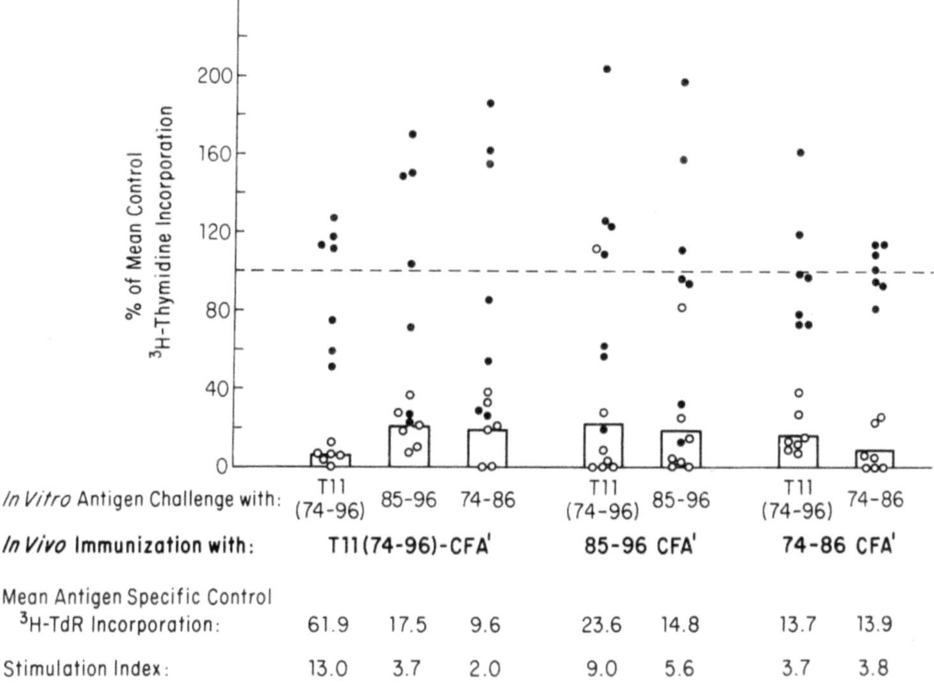

Mean Antigen Specific Control ^{3}H-TdR Incorporation:	61.9	17.5	9.6	23.6	14.8	13.7	13.9
Stimulation Index:	13.0	3.7	2.0	9.0	5.6	3.7	3.8

Fig. 2. Response of T11-neonatally tolerized B10.A mice to challenge with T11, 74–86, and 85–96. Tolerance was induced by injection of T11 in FIA at 24–48 h after birth and assessed by challenge with peptide in FCA at 8–10 weeks of age. The in vitro proliferative response to 7 μM T11, 74–86, and 85–96 was measured. The data are expressed as percentage of the response of normal control mice except for challenge with 85–96 where the control group were 74–86-tolerant B10.A mice. Each *filled circle* corresponds to a single nontolerized control; the *open circles* to each of the tolerant mice

with T11 in FCA. The data in Fig. 2 demonstrate that these animals do not respond to T11: thus, small HEL peptides can induce tolerance.

The T11-tolerant mice were also challenged with 74–86 and 85–96 to test for latent reactivity, but no response was observed (Fig. 2). This result is consistent with the failure of challenge with the peptides L2 and L3 in CFA to stimulate responses in HEL-induced neonatal tolerance and suggests that tolerance to T11 is caused by clonal inactivation. If T11 tolerance were due to suppression, peptides 74–86 and 85–96 would have to contain a Ts-inducing determinant in addition to a Tp-inducing determinant, which would seem unlikely for a 13 or 12 amino acid peptide. Nevertheless, this possibility was directly tested by inducing tolerance with 74–86 and later challenging with T11. If clonal inactivation had occurred, it would only have affected T cells which recognized determinants on the tolerogen and not those recognizing adjacent determinants. If Ts on 74–86 were truly involved, they should act on the response to *both* determinants on T11 in a similar manner to that in which the HEL N-terminal-specific Ts suppress the T-cell proliferative response to determinants on other

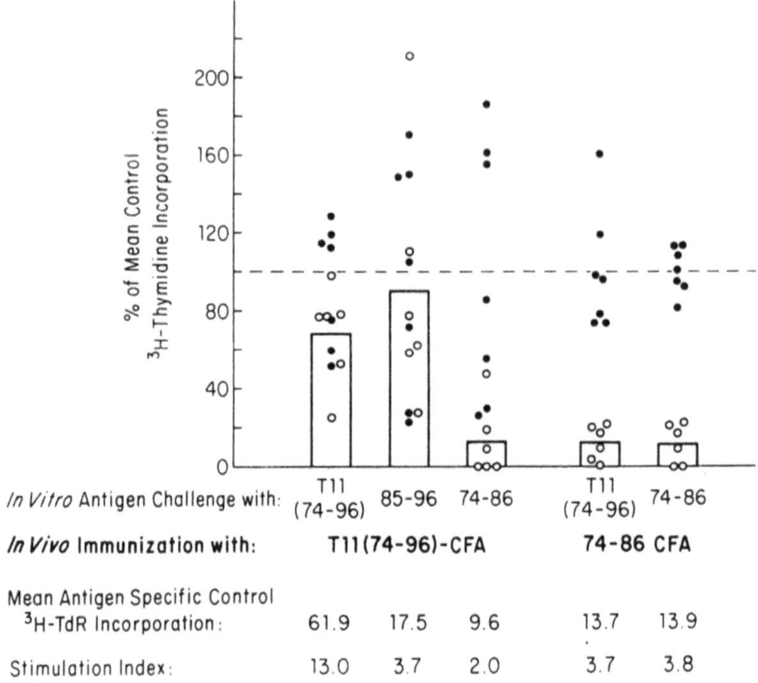

Fig. 3. Response of 74–86-neonatal tolerant B10.A mice to challenge with T11 and 74–86. The data are expressed as in Fig. 2

regions of the lysozyme molecule. Therefore, challenge of 74–86-tolerant mice with T11 should fail to induce a response to either determinant due to dominant suppression. The data given in Fig. 3 demonstrate that the first possibility is correct. Immunization of 74–86-tolerant mice with T11 does induce a response and this response is limited to 85–96. Therefore, tolerance is specific for epitopes on the tolerogen and the response to nearby linked determinants is unaffected. These data support the clonal inactivation model and argue against the model of active suppression.

It has been reported that tolerance to minor histocompatibility antigens is restricted to the host animal's MHC gene complex, i.e., the animal is only tolerant to a particular antigen when presented by self antigen-presenting cells (GROVES and SINGER 1983; MATZINGER et al. 1984; RAMMENSEE and BEVAN 1984). This suggests that tolerance is induced by antigen presented in the context of MHC gene products and that induction of tolerance is similar to antigen presentation during stimulation of a response. Assuming that tolerance were the result of clonal inactivation, any peptide which stimulates a response should be able to induce tolerance and in fact immunogenic and tolerogenic peptides should be identical. To confirm this prediction the 74–86-induced response was analyzed and smaller peptides synthesized to define the minimal size for an immunogenic peptide in this system. It was found that 74–82 could recall a response in 74–86-primed lymph node cells, while peptide 77–86 did not stimu-

Table 2. Response to HEL peptide 74–86 after neonatal injection with 74–82 and 77–86

Tolerogen	Number of mice	[^{3}H]-Thymidine incorporation per culture (Δ cpm $\times 10^{-3}$)		
		Medium	PPD	74–86
–	7	(3.7 ± 2.2)	137.4 ± 31.5	20.2 ± 16.2
74–82	8	(2.5 ± 1.2)	157.3 ± 29.3	4.7 ± 5.7
77–86	7	(2.0 ± 0.5)	142.8 ± 22.7	31.3 ± 17.8

Tolerance was induced by injection of 7 nmol of peptide in FIA at 24 h and 72 h after birth. The mice were challenged with 7 nmol of peptide 74–86 in FCA at 8 weeks and 10 days later popliteal and inguinal lymph nodes were removed. Lymph node cells were added at 4×10^5 per well and cultured with medium alone, or with Purified Protein Derivative (of tuberculin) at 5 µg/ml and peptide 74–86 at 14 µM. Incorporation of 1 µCi[^{3}H]-thymidine was measured during the last 18 h of a 5-day culture. The data are the arithmetic mean of cpm $\times 10^{-3}$ of the individual mice ± standard deviation

late any significant degree of proliferation. To test whether either of these peptides could induce tolerance, neonatal B10.A mice were injected with either 74–82 or 77–86. These animals were challenged at age 8 weeks with 74–86 in FCA and it was discovered that the immunogenic peptide 74–82 could induce tolerance but the overlapping peptide 77–86 did not (Table 2). Peptide 74–82 is only nine amino acids long and is close to the minimal size reported for a peptide to induce a response. It would be highly improbable that this peptide also contain a Ts-inducing determinant and so this is a further argument against a role for Ts in tolerance induction. Furthermore, these results suggest that any immunogenic peptide has the capacity to induce tolerance.

4 Neonatal Tolerance to Cytochrome *c* Peptides: Analysis of the Specificity of Tolerance

We now sought to focus on the exact specificity of tolerance induction to a minimal, immunogenic peptide. As an ideal starting point, we looked for a situation with a small peptide where a single amino acid change would disrupt its epitope, so that the specificity of tolerance could be ascertained. At the same time, the identity of the rest of the peptide could be used to rule out suppressive mechanisms. Such a case was provided by using synthetic cytochrome *c* peptides corresponding to amino acids 93–103 of moth cytochrome *c* in neonatal B10.A mice (GAMMON et al. 1986).

The immune response to cytochrome *c* has been extensively studied and the Tp-inducing determinants characterized (SOLINGER et al. 1979; HANSBURG et al. 1983). The major determinant is located at the carboxy-terminal end of the molecule and is contained within the peptide 93–103. It has been shown that amino acid residue 99 constitutes a critical part of the T-cell epitope, so that substitutions at this site radically alter the composition of the responding population. Thus the responses to these substituted peptides are almost entirely

Table 3. Specificity of response to cytochrome c peptides

Immunogen	[³H]-Thymidine incorporation per culture (Δcpm $\times 10^{-3}$)		
	K 99	Q 99	R 99
K 99	104.2 ± 9.8	3.9 ± 1.4	3.9 ± 0.7
Q 99	3.5 ± 0.7	66.6 ± 4.0	4.1 ± 3.7
R 99	4.9 ± 1.4	1.1 ± 0.7	51.8 ± 5.3

The popliteal and inguinal lymph node cells from three 8–10 week old primed B10.A mice were pooled and added to wells with a 10 µM concentration of each cytochrome peptide. Incorporation of [³H]-thymidine for the last 18 h of a 5-day in vitro culture was measured

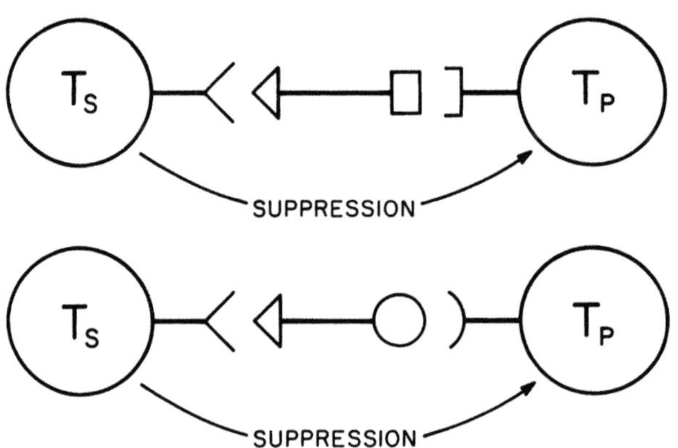

Fig. 4. Shared suppressor determinant should lead to cross-tolerance of unrelated Tp

noncross-reactive with each other. Three peptides were synthesized: K 99, with a lysine at position 99 (which is equivalent to 93–103 of the moth sequence), Q 99, and R 99, with the single amino acid substitution of glutamine or arginine for lysine respectively at residue 99. Each peptide stimulates an essentially unique noncross-reactive T-cell population as shown in Table 3. If tolerance were due to clonal inactivation, then the same cells normally stimulated by a peptide would be inactivated. Hence the specificity of tolerance would match that of the response and each cytochrome peptide would induce tolerance for itself only. If tolerance were due to suppression, cross-tolerance between the peptides would be expected because the important factor in suppressor-mediated tolerance would be the sharing of a common suppressor determinant by the cytochrome peptides (Fig. 4). The three cytochrome peptides are identical except for a single substitution in the Tp determinant and, since Ts- and Tp-inducing determinants appear to be distinct, this substitution would not be expected to affect a coexistent Ts-inducing determinant.

Injection of cytochrome peptides into neonatal mice induced profound tolerance upon challenge at 8 weeks (Table 4). Tolerance was specific to the peptide used for its induction. Thus, K 99-tolerant mice did not respond to challenge

Table 4. Specificity of neonatal tolerance to cytochrome c peptides

Mice	[³H]-Thymidine incorporation per culture (Δ cpm $\times 10^{-3}$)		
	K 99	Q 99	R 99
Normal	107.7	54.7	68.0
K 99 Tolerant	<u>14.0</u>	90.2	65.9
R 99 Tolerant	80.6	50.7	<u>5.6</u>

Tolerance was induced as described in Table 2. The in vitro proliferative response to each cytochrome peptide following in vivo challenge with the same peptide in FCA was measured by [³H]-thymi-
dine incorporation. The in vitro peptide concentration was 10 µM.
The underlined values represent cases of tolerance induction (significant differences in proliferation obtained compared to untreated
mice)

with K 99 in FCA, but did respond to both R 99 and Q 99. Therefore, the
specificity of tolerance matches that of the normal response and there is no
sign of cross-tolerance. This is strong evidence for a mechanism of neonatal
clonal inactivation and argues against active suppression in neonatal tolerance
to small immunogenic peptides.

5 Conclusions

Several arguments can be drawn from the data presented above which support
the hypothesis that neonatal tolerance to immunogenic peptides is due to clonal
inactivation. No latent response was found in T11-tolerant mice upon challenge
with 74–86 and 85–96. Tolerance was shown to be restricted to epitopes on
the tolerogen, as T11 could stimulate a response (specific for 85–96) in 74–86-tol-
erant mice. Tolerance was induced even by the minimal immunogenic peptide
74–82 and it appears that any peptide which stimulates a response can induce
tolerance. Finally the specificity of tolerance to different cytochrome c peptides
was demonstrated to match that of the normal response to each peptide and
no sign of suppressor-mediated cross-tolerance was detected.

Putative idiotypic mechanisms which would account for complete tolerance
to a peptide seem unlikely, although they have not been excluded in our work.
Activation of the group of Tp or Th specific for a minimal peptide might
be followed by induction of a set of Ts which recognize the receptors on the
Tp/Th and regulate their activity. Recent work in this laboratory (SHASTRI
et al. 1985) has indicated that a heterogenous array of clones arises after immunization with a peptide, and it is evident (GOVERMAN et al. 1985) that these clones
do not use the same V_β genes. In this case, regulation by anti-idiotypic Ts
would require a large battalion of Ts, one for each target idiotype (Id) unless
there were significant sharing of a predominant Id: exploring these possibilities
remains an important goal of our tolerance studies with peptides.

In these experiments, tolerance was induced in neonatal animals, and so it must be considered whether the same process can occur in adults. Although we have no evidence that clonal inactivation can occur in older mice, there are many reports of adult-induced tolerance to proteins and haptens in which tolerance appears to be due to this mechanism (e.g. DOYLE et al. 1976; MILLER et al. 1979; LOWY et al. 1984). Furthermore, T-cell clones can be directly inactivated in vitro by peptides, in the absence of other T-cell subpopulations (LAMB et al. 1983; LEVICH et al. 1985). For example, tolerance can be induced in human T-cell clones specific for influenza peptides by exposure to a high concentration of antigen. Unresponsiveness lasts for longer than 1 week and does not involve death of the antigen-specific cells. Induction of in vitro tolerance requires the presence of class II MHC gene products and can be blocked by the appropriate anti-class II monoclonal antibodies (LAMB and FELDMANN 1984). This suggests that clonal inactivation may occur even in adult mice, and so all peptides should be considered potentially tolerogenic. Perhaps those clones with the highest affinity for antigen/MHC will be those with the lowest threshold for tolerance induction. Therefore, factors such as dosage, route of administration, and use of adjuvants are important in stimulating a strong response and avoiding tolerance. However, in addition to clonal inactivation, the induction of Ts may cause unresponsiveness, and the ability of peptides to induce suppression must be examined.

There are several examples of Ts induction by peptides. NC, an HEL peptide corresponding to amino acid residues 1–17 and 120–129 linked by a disulfide bond 6–127, can stimulate Ts in B6 mice and suppress the anti-HEL plaque forming cell response (ADORINI et al. 1979). The delayed type hypersensitivity reaction to HEL in A/J mice is suppressed by prior injection of the disulfide-linked peptide 29–54: 109–123 (SEMMA et al. 1981). Although it is clear that peptides can induce Ts, the structural requirements of a suppressogenic peptide are unknown. In particular, it is still debated whether Ts recognize native or processed antigen and so it is possible that suppressogenic peptides may have to have a stable conformation similar to that of the peptide when lodged within the native molecule (GOODMAN and SERCARZ 1983). Of special importance to our considerations is whether a suppressogenic peptide will induce a state of suppressor memory which will be expressed for the peptide itself as well as any immunogenic determinants attached to it.

Ts induction by peptides is not necessarily detrimental. There are several instances, e.g., in allergy and autoimmune disease, where stimulation of suppression would be advantageous. These Ts would act in different ways. First, using synthetic peptides it should be possible to stimulate Ts which cross-react with the native protein in the host. Second, an SD unrelated to the protein could be coupled to it so that the Ts would then act on T cells responding to the host protein. An example of the latter "suppressor-determinant-grafting" approach is the use of TNP-specific Ts to suppress kidney allograft rejection in rats (HUTCHINSON et al. 1985). In this experiment, TNP-specific Ts were transferred into low dose-irradiated recipients with TNP-modified alloantigens one day before renal transplantation. If the graft bore the same alloantigens as those injected with the TNP-specific Ts graft survival was substantially prolonged.

References

Adorini L, Harvey MA, Miller A, Sercarz EE (1979) The fine specificity of regulatory T cells. II. Suppressor and helper T cells are induced by different regions of hen egg-white lysozyme (HEL) in a genetic nonresponder mouse strain. J Exp Med 150:293

Berkower I, Matis LA, Buckenmeyer GK, Gurd FRN, Longo DL, Berzofsky JA (1984) Identification of distinct predominant epitopes recognized by myoglobin-specific T cells under the control of different Ir genes and characterization of representative T cell clones. J Immunol 132:1370

Bixler GS, Yoshida TS, Atassi MZ (1985) Antigen presentation of lysozyme: T-cell recognition of peptide and intact protein after priming with synthetic overlapping peptides comprising the entire protein chain. Immunology 56:103

Doyle MV, Parks DE, Weigle WO (1976) Specific suppression of the immune response by HGG tolerant spleen cells. 1. Parameters affecting the level of suppression. J Immunol 116:1640

Gammon G, Dunn KD, Shastri N, Oki A, Wilbur S, Sercarz EE (1986) Neonatal T-cell tolerance to minimal immunogenic peptides is caused by clonal inactivation. Nature 319:413

Germain R (1981) Accessory cell stimulation of T cell proliferation requires active antigen processing, I a-restricted antigen presentation, and a separate non-specific second signal. J Immunol 127:1964

Goodman JW, Sercarz EE (1983) The complexity of structures involved in T cell activation. Ann Rev Immunol 1:465

Goverman J, Minard K, Shastri N, Hunkapiller T, Hansburg D, Sercarz EE, Hood L (1985) Rearranged β T-cell receptor genes in a helper T-cell clone specific for lysozyme: No correlation between V_β gene segments and antigen specificity or MHC restriction. Cell 40:859

Groves ES, Singer A (1983) Role of the H-2 complex in the induction of T cell tolerance to self minor histocompatibility antigens. J Exp Med 158:1483

Hansburg D, Fairwell T, Schwartz RH, Appella E (1983) The T lymphocyte response to cytochrome c. IV. Distinguishable sites on a peptide Ag which affect antigenic strength and memory. J Immunol 131:319

Hutchinson IV, Barber WH, Morris PJ (1985) Specific suppression of allograft rejection by trinitrophenyl (TNP)-induced suppressor cells in recipients treated with TNP-haptenated donor alloantigens. J Exp Med 162:1409

Jensen PE, Kapp JA (1985) Genetics of insulin-specific helper and suppressor T cells in non-responder mice. J Immunol 135:2990

Krzych U, Fowler A, Sercarz EE (1983) Antigen structures used by regulatory T cells in the interaction among T suppressor, T helper and B cells. In: Celada F, Schumaker VN, Sercarz EE (eds) Protein Conformation as an Immunological Signal. Plenum, London, p 395

Lamb JR, Feldmann M (1984) Essential requirement for major histocompatibility complex recognition in T-cell tolerance induction. Nature 308:72

Lamb JR, Skidmore BJ, Green N, Chiller JM, Feldmann M (1983) Induction of tolerance in influenza virus-immune T lymphocyte clones with synthetic peptides of influenza hemagglutinin. J Exp Med 157:1434

Levich JD, Parks DE, Weigle WO (1985) Tolerance induction in antigen-specific helper T cell clones and lines in vitro. J Immunol 135:873

Loblay RH, Fazekas de St. Groth B, Pritchard-Briscoe H, Basten A (1983) Suppressor T cell memory. II. The role of memory suppressor T cells in tolerance to human gamma globulin. J Exp Med 157:957

Lowy A, Drebin JA, Monroe JG, Granstein RD, Greene MI (1984) Genetically restricted antigen presentation for immunological tolerance and suppression. Nature 308:373

Manca F, Clarke JA, Miller A, Sercarz EE, Shastri N (1984) A limited region within hen eggwhite lysozyme serves as the focus for a diversity of T cell clones. J Immunol 133:2075

Matzinger P, Zamoyska R, Waldmann H (1984) Self tolerance is H-2 restricted. Nature 308:738

Miller SD, Wetzig RP, Claman HN (1979) The induction of cell-mediated immunity and tolerance with protein antigens coupled to syngeneic lymphoid cells. J Exp Med 149:758

Oki AN, Sercarz EE (1985) T cell tolerance studied at the level of antigenic determinants. 1. Latent reactivity to lysozyme peptides that lack suppressogenic epitopes can be revealed in lysozyme tolerant mice. J Exp Med 161:897

Rammensee HG, Bevan MJ (1984) Evidence from in vitro studies that tolerance to self antigens is MHC-restricted. Nature 308:741

Rosenwasser LJ, Barcinski MA, Schwartz RH, Rosenthal AS (1979) Immune response gene control of determinant selection. II. Genetic control of the murine T lymphocyte proliferative response to insulin. J Immunol 123:471

Semma M, Sakato N, Fujio H, Amano T (1981) Regulation of delayed-type hypersensitivity (DTH) response to hen egg-white lysozyme (HEL): Induction of suppressor T-cells with soluble HEL derivative peptide in A/J mice. Immunol Letters 3:57

Servis C, Seckler R, Nagy ZA, Klein J (1986) Two adjacent epitopes on a synthetic dodecapeptide induce lactate dehydrogenase B-specific helper and suppressor T cells. Proc Roy Soc B

Shastri N, Oki A, Miller A, Sercarz EE (1985) Distinct recognition phenotypes exist for T cell clones specific for small peptide regions of proteins. Implications for the mechanism underlying major histocompatibility complex-restricted antigen recognition and clonal deletion models of immune response gene defects. J Exp Med 162:332

Shimonkevitz R, Kappler J, Marrack P, Grey HM (1983) Antigen recognition by H-2-restricted T cells. I. Cell-free antigen processing. J Exp Med 158:303

Solinger AM, Ultee ME, Margoliash E, Schwartz RH (1979) T lymphocyte response to cytochrome c. I. Demonstration of a T-cell heteroclitic proliferative response and identification of a topographic antigenic determinant on pigeon cytochrome c whose immune recognition requires 2 complementing immune response gene. J Exp Med 150:830

Waters RH, Pilarski LM, Wegmann TG, Diener E (1979) Tolerance induction during ontogeny. 1. Presence of active suppression in mice rendered tolerant to human-γ-globulin in utero correlated with the breakdown of the tolerant state. J Exp Med 149:1134

Weigle WO (1980) Analysis of autoimmunity through experimental models of thyroiditis and allergic encephalomyelitis. Adv Immunol 30:159

Ziegler K, Unanue ER (1981) Identification of a macrophage antigen-processing event required for I-region-restricted antigen presentation to T lymphocytes. J Immunol 127:1869

Immune Response to Synthetic Herpes Simplex Virus Peptides: The Feasibility of a Synthetic Vaccine

E. HEBER-KATZ and B. DIETZSCHOLD

1 Introduction

Synthetic peptides which are selected from immunogenic proteins offer an approach to the development of a vaccine with optimal properties. The strategy used for the design of such a synthetic vaccine includes: (1) the identification of regions in the protein molecule which act as determinants for B cells as well as for T cells; (2) the chemical synthesis of identified antigenic determinants; and (3) the attachment of the synthetic peptides to suitable carriers which allow optimal presentation of the peptide antigen. For our studies we have chosen glycoprotein D(gD) of the herpes simplex virus (HSV). This antigen appears to be the major target of the immune response to HSV (PAOLETTI et al. 1984; CHAN 1983; LASKY et al. 1984; LONG et al. 1984; SPEAR 1984) and therefore represents a logical choice for a subunit vaccine against an HSV infection. We will describe the localization and chemical synthesis as well as the physicochemical and immunological characterizaton of a major antigenic site of HSV gD. Synthetic peptides comprising this antigenic site are recognized by HSV type-common and type-specific antibodies, as well as by HSV-primed T cells, and immunization with these molecules induces both a B cell and a T cell response. Most important is the finding that a single injection of these peptides can confer long-term protection against a lethal HSV infection, but only when the peptides are incorporated into liposomes. The exact nature of this immune and protective response will be the subject of this paper.

The Wistar Institute, 36th Street at Spruce, Philadelphia, PA 19104, USA

Current Topics in Microbiology and Immunology, Vol. 130
© Springer-Verlag Berlin · Heidelberg 1986

2 Defining Antigenic Sites on the HSV gD Glycoprotein

A panel of monoclonal antibodies was used to define eight different epitopes on gD (EISENBERG et al. 1982a), some of which were common to both gD-1 and gD-2. The type common determinants, as illustrated by group VII monoclonal antibodies, reacted with both gD-1 and gD-2, exhibited type-common virus neutralizing activity, and were able to recognize a reduced and alkylated form of gD, strongly suggesting that this type-common epitope was sequential in nature. On the other hand, the type-specific determinants examined were found on regions of the molecule which had significant differences in amino acid sequence between gD-1 and gD-2 (WATSON et al. 1982; WATSON 1983; LASKY and DOWBENKO 1984). They appear to reflect differences in structure as they are conformation dependent and not maintained on denatured gD.

The localization of the type-common determinant defined by group VII antibodies was carried out by using *Staphylococcus aureus* protease V8 to isolate a 12000-dalton fragment of gD-1 (EISENBERG et al. 1982b; COHEN et al. 1984). This fragment was bound by the group VII monoclonal antibody 170. Tryptic peptide analysis indicated the presence of an arginine-containing peptide called "f" in this 12000-dalton fragment. The f peptide was isolated and its size determined on Bio-Gel P4 to be 700 daltons. When the amino acid sequence of the peptide was compared to that obtained from the nucleotide sequence of gD, it appeared that the f peptide coincided with residues 36–41. Taken together with the fact that a leader sequence included residues 1–23, this suggested that the group VII determinant was near the amino terminus of gD-1.

It has been shown that gD is structurally and antigenically similar, though not identical, in the two serotypes of HSV (HSV-1 and HSV-2) (BALACHANDRAN

Table 1. Primary sequences and helix contents of synthetic peptides mapped within the first 23 amino acids of HSV gD

Residue	Amino acid sequence	Helix content (%)[a]
1–23(1)	K Y A L A D A S L K M A D P N R F R G K D L P	17
1–23(2)	K – – – – – P – – – – – – – – – – – – – N – P	<5
1–23(H)	K – – – – – P – – – – – – – – – – – – – D – P	5
1–16(1)	K – – – – – A – – – – – – – – R	17
1–16(2)	K – – – – – P – – – – – – – – R	15
8–23(1)	S L K M A D P N R F R G K D L P	<10
8–23(2)	S – – – – – – – – – – – – N – P	<5
12–23(1)	A – – – – – – – – D – P	nd
12–23(2)	A – – – – – – – – N – P	nd
17–23(1)	F – – – N – P	[b]
17–23(2)	F – – – N – P	[b]

[a] Helix content of the fragment in TFE derived from the spectra by the method of Greenfield and Fasman (GREENFIELD and FASMAN 1969)
[b] Molecules too small for calculation of helicity

nd, not determined

et al. 1982; EISENBERG et al. 1982a; EISENBERG et al. 1980; SHOWALTER et al. 1981). If one compares the sequence of gD-1 (WATSON et al. 1982) and gD-2 (WATSON 1983; LASKY and DOWBENKO 1984), differences are found in two residues, 7 and 21, within the first 23 N-terminal amino acids: Ala-7 and Asp-21 in gD-1, and Pro-7 and Asn-21 in gD-2.

To define further the epitopes within the group VII determinant or antigenic region recognized by both type-common and type-specific antibodies, a series of overlapping peptides ranging in length from 7 to 23 amino acids and corresponding to the amino acid residues 1–23 of gD-1 and of gD-2 was synthesized using Merrifield solid-phase methods (ERICKSON and MERRIFIELD 1976; Table 1). Peptides 1-23(H) and 3-23(H) were synthesized to represent "hybrids" having the type 2 specific residue (Pro: P) in position 7 and the type 1 specific residue (Asp: D) in position 21.

3 Secondary Structure of Synthetic Peptides

To examine the effects of the amino acid changes at positions 7 and 21 on the secondary structure of peptides 1–23(1), 1–23(2), and 1–23(H), computer analyses based on the predictive rules of CHOU and FASMAN (1978) were carried out. It was shown that the 1–23(1) peptide consisted of an alpha helical structure at the N-terminus followed by one beta turn (Fig. 1). When the alanine at position 7 was substituted by a proline, the alpha helical structure at the N-terminus was abolished and an additional beta turn was predicted (1–23(H); Fig. 1). The peptide 1–23(2) with asparagine at position 21 was predicted to have an additional beta turn at the C-terminal end, yielding a structure consisting of three beta turns (Fig. 1; DIETZSCHOLD et al. 1984).

Besides predictive computer analysis, circular dichroism (CD) measurements were taken in both an aqueous solution and trifluorethanol (TFE) to determine whether the synthetic fragments in fact formed ordered structures under appropriate conditions (HEBER-KATZ et al. 1985). The peptide fragments in an aqueous solution showed characteristics of unordered polypeptides, i.e., strong negative bands below 200 nm. If CD measurements were carried out in the helix-promoting solvent TFE, a helical-type spectrum was revealed (a broad shoulder at about 220 nm, a negative band around 204 nm, and a positive band near 190 nm).

From this data the highest helix content (15%–20%) (see Table 1) was calculated for fragments 1–23(1) and 1–16(1). The calculated helicity accords well with the secondary structure predictions. However, the C-terminal fragments 8–23(1) and 8–23(2) and both 1–23 Pro-7 fragments, 1–23(H) and 1–23(2), showed a small helix content (around 5%). On the basis of the spectral characteristics of these peptides the dominance of unordered conformations with a strong negative band below 200 nm could also be excluded (HEBER-KATZ et al. 1985).

Predicted Secondary Structure of HSV gD-peptides

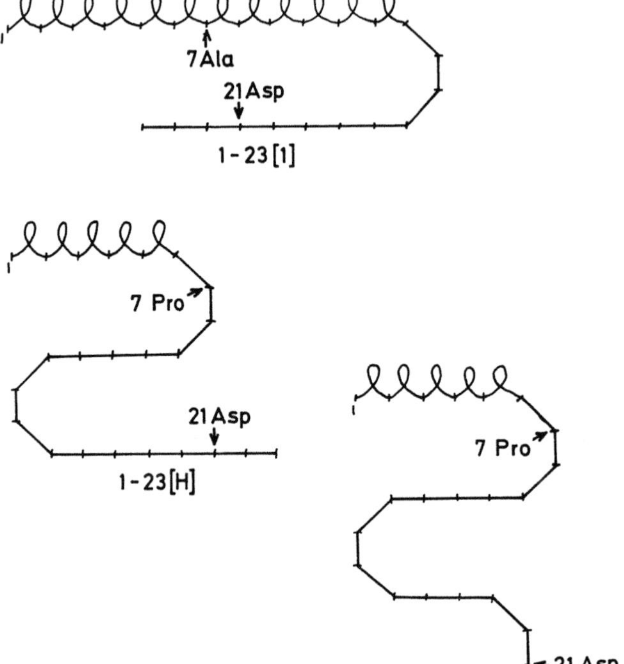

Fig. 1. Schematic diagram of the predicted secondary structure potentials. Each residue is represented in helical (ꝉ), beta sheet (↦), and coil conformations ( ). Shown are the 1–23(1), 1–23(H), and 1–23(2) peptides

4 Antibody Responses

The peptides were tested for their ability to bind anti-gD-specific polyclonal and monoclonal antibodies (COHEN et al. 1984; DIETZSCHOLD et al. 1984). An immunoblot assay showed that a rabbit anti-gD-1 hyperimmune serum reacted with gD-1 and gD-2 peptides, including 1–23, 3–23, 8–23, and 12–23 but not 1–16 or 17–23. Thus, anti-gD-1 serum did not distinguish between aspartic acid and asparagine at position 21 and did not detect any determinant in the N-terminal region. The same reactivity pattern was seen with a mouse monoclonal anti-gD-1.

Rabbit anti-gD-2 hyperimmune serum, on the other hand, appeared to be extremely type-specific, reacting with the type 2 peptides except for the smallest, 17–23(2), but not with the type 1-specific peptides. It did react with the hybrid peptides, 1–23(H) and 3–23(H), which have the gD-2 specific Pro at position

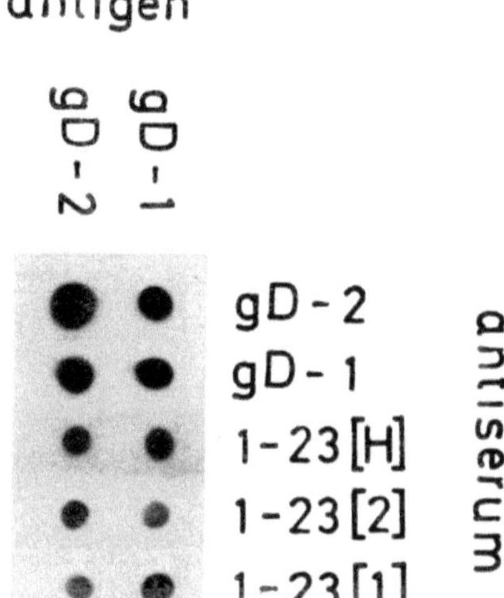

Fig. 2. Dot blot analysis of antisera from animals immunized with gD or the peptides, and their binding activity on gD-1 and gD-2

7. Moreover, anti-gD-2 serum recognized 1–16(2) but not 1–16(1), in contrast to the anti-gD-1 serum which recognized neither.

The results of the binding studies indicate the presence of two regions of gD, within the amino acids 1–23 capable of binding type-specific anti-gD-2 antibodies. One of these (region I) is contained within the first 16 amino acids of gD-2 and appears to involve proline at position 7 and one or both of the lysines at positions 1 and 10, since citraconylation destroys binding activity. The same peptide with Ala in position 7 is not recognized by anti-gD-1 or anti-gD-2 sera, possibly because this sequence is buried within the intact glyco-protein. The second region (region II), which denotes type 2 specificity, is located closer to the carboxy-end of the 1–23 sequence. In the case of anti-gD-2 sera, a reaction was seen with peptides containing 8–23 or 12–23, but only when residue 21 is asparagine.

Another epitope within region II could be defined by the monoclonal anti-body 170 which could bind peptides 8–23(1) and 8–23(2), suggesting that this epitope is unaffected by a change of the amino acid at 21. Since the antibody could react with the citraconylated 1–23(H) peptide, it is likely that the 170 epitope does not involve the two lysines located at positions 10 and 20, and this epitope is probably located between residues 11–19, where there is complete homology between gD-1 and gD-2, and is predicted to assume a similar second-ary structure. Thus, the 170 epitope is truly type-common.

To determine whether these synthetic peptides could mimic the immunogenic properties of the entire protein, the 1–23 peptides were coupled to KLH, and rabbits were injected multiple times. Antisera produced against these peptides reacted strongly with either intact gD-1 or gD-2 (Fig. 2). Furthermore, these antisera were found to neutralize both HSV-1 and HSV-2.

When the shorter 1–16 or 8–23 peptides were used, there was reactivity against the peptides; however, there was little or no reactivity against the native

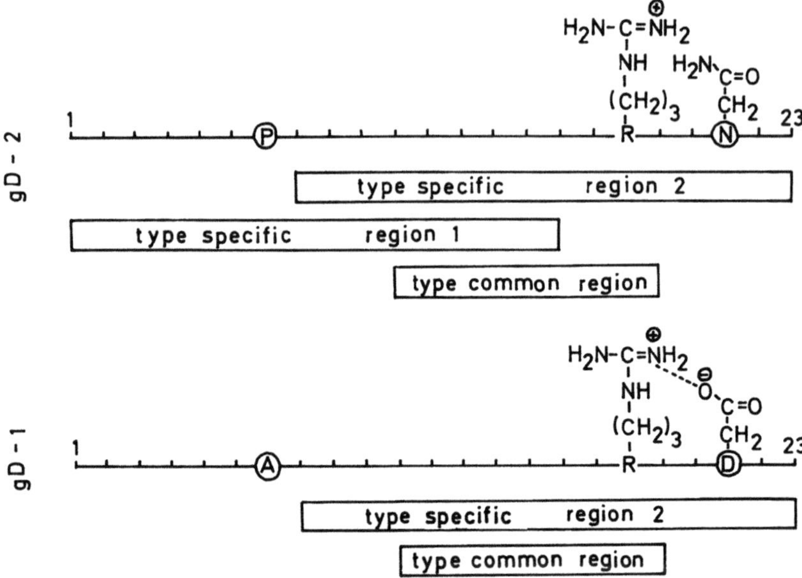

Fig. 3. Diagrammatic representation of the N-terminal 1–23(1) and 1–23(2) peptides corresponding to gD-1 and gD-2, showing the type-common and type-specific antibody binding regions and demonstrating a possible stabilizing interaction in the gD-1 peptide (HEBER-KATZ et al. 1985)

glycoproteins, indicating that regions I and II are presented differently on the short peptides, the 1–23 peptides, and the native glycoproteins. The poor reactivity of the 1–16 and 8–23 peptide antibodies may lie in the greater conformational freedom of these peptides in solution, whereas the larger 1–23 peptides, which induce antibody cross-reactions with the native glycoprotein, have a more restricted conformational freedom.

Thus, we imagine the B-cell determinants present on the 1–23(1) and 1–23(2) peptides as shown in Fig. 3.

5 T-Cell Responses

Besides examining the ability of these peptides to induce a B-cell antibody response, we also examined the T-cell response. We found that immunizing mice with any of the 1–23 peptides resulted in a potent, proliferative T-cell response in vitro to all but the shortest peptides. The fine specificity of this response had similarities to the B-cell response; the T-cell specificities generated by these peptides mapped again to two major regions of the 1–23 peptides, one which could be included within residues 1–16 and a site included within residues 8–23. Thus, employing the same peptides that were used to determine the antibody sites, we found that these T-cell and B-cell determinants overlap. As seen in Fig. 4, immunization with different peptides utilizes these different

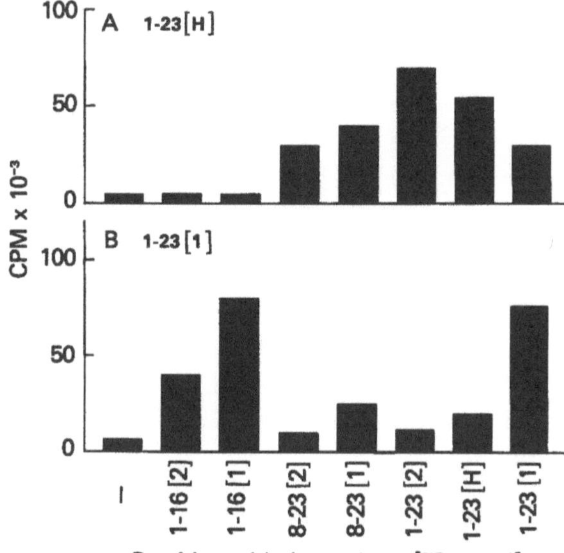

Fig. 4. Response pattern of B10.A T-cell lines to synthetic HSV peptides. Mice were immunized with 0.1 mg of either 1–23(H) or 1–23(1) peptide, and their T cells were tested for proliferation in vitro using 50 µg/ml of the peptides shown by ^{3}H-thymidine incorporation

sites. Thus, immunization with the 1–23(1) sequence primes mainly for a 1–16 response and 1–23(H) primes for an 8–23 sequence response.

We found several interesting relationships associated with these different T-cell sites which relate to both the secondary structure of the peptides and the ability of clones to recognize antigen in association with a given Ia isotype.

5.1 Structure–Function Relationship

We found that the reactivity patterns of peptide-primed lymph node T cells appeared to correlate with the secondary structure of the peptides (HEBER-KATZ et al. 1985). The region I peptides (1–16), according to CD measurements, are alpha helical and the region II peptides (8–23) are nonhelical. When we immunized with a helical peptide, 1–23(1), the response was directed to the helical determinants 1–23(1), 1–16(1), and 1–16(2) (Fig. 4). On the other hand, immunization with a nonhelical peptide, 1–23(H), resulted in a response directed to the nonhelical peptides, 1–23(H), 1–23(2), 8–23(1), and 8–23(2). The structure of the peptide was not merely coincidental with the T-cell response as we found that when the helical peptide 1–16(2) was located in a larger molecule that was nonhelical, 1–23(H) and 1–23(2), there was no "helix-primed" response. Thus, 1–23(1)-specific T cells did not respond to these molecules. In the reverse situation, a nonhelical peptide, 8–23, when situated within a larger molecule with helical structure, 1–23(1), induced a poor "nonhelix-primed" response. Thus, 1–23(H)-primed T cells responded poorly to 1–23(1). These results were further confirmed with antigen-specific T-cell hybridoma clones.

We feel that these data support the idea that secondary structure is important for T-cell recognition of antigen and also indicate that there is not a preference

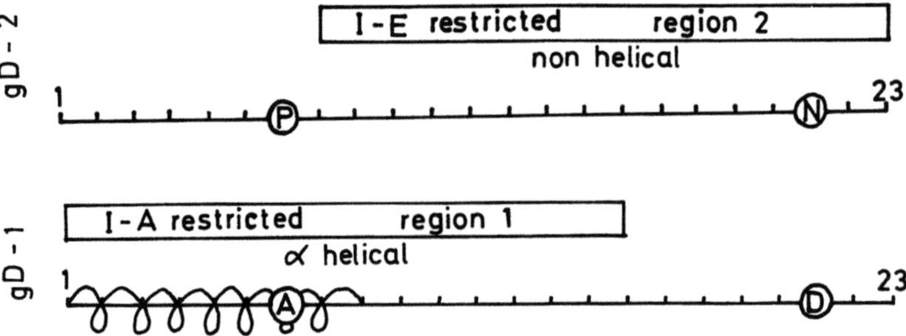

Fig. 5. Diagrammatic representation of the 1–23 peptides showing T-cell sites and their structural and MHC restriction associations

for alpha helical structures. Furthermore, it appears that amino acid substitutions made downstream from the T-cell determinant can affect T-cell responses, a statement which may be important in trying to map specific contact residues in T cell-antigen-Ia interactions.

5.2 MHC Restriction

We also found that in mapping the T-cell sites on the 1–23 peptides, the region I(1–16) site was presented exclusively in association with the I-A molecule and the region II (8–23) site was presented by the I-E molecule (Fig. 5). Thus, all T-cell clones isolated with a given specificity responded with a particular Ia molecule. Furthermore, mice which only expressed the I-A molecule could only respond to the 1–16 peptides of region I and not the 8–23 peptides of region II. We cannot relate the secondary structure of these peptides to the particular Ia molecule used; however, given the accumulating data indicating a physical interaction between antigen and Ia, we may find an important association between antigen structure and MHC (Pincus et al. 1983; DeLisi and Berzofsky 1985; Babbitt et al. 1985).

Upon further examination of the I-E restricted 8–23 directed response, we found that it was highly MHC degenerate in a manner similar to that seen with cytochrome c (Heber-Katz et al. 1982). Furthermore, allogeneic antigen-presenting cells (APC) presented antigen in a haplotype-specific manner (Fig. 6). This APC-directed change in antigen specificity supported the notion that Ia and this antigen interact, and that the peptide has two functional sites, and it extends our previous observations (Heber-Katz et al. 1983) on determinant selection to viral determinants of biological significance. The functional mapping of these sites by selective amino acid substitutions is involved with concomitant changes in spatial structure. Thus, we were concerned that amino acid substitutions with subsequent changes in specificity may not be mapping actual intermolecular contact points. In fact, as previously discussed, we found that amino acid substitutions just *outside* the determinants had significant effects on fine specificity. CD measurements on these peptides showed that changes in specificity correlated strongly with changes in secondary structure.

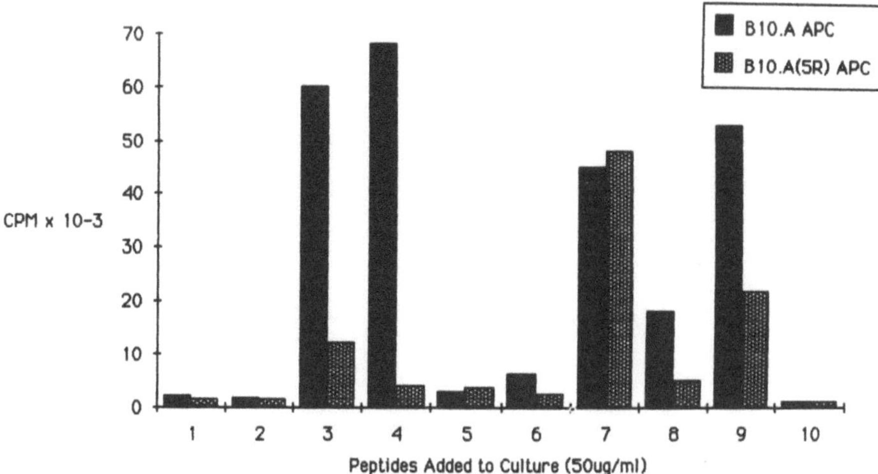

Fig. 6. A T-cell hybridoma derived from a B10.A animal immunized to the 1–23(H) peptide and stimulated in culture in the presence of B10.A or B10.A(5R) spleen with 50 μg/ml of the following peptides: 1, 17–23(2); 2, 17–23(1); 3, 8–23(2); 4, 8–23(1); 5, 1–16(2); 6, 1–16(1); 7, 1–23(2); 8, 1–23(1); 9, 1–23(H); and 10, no antigen

6 Protection Against a Lethal HSV Infection by a Synthetic Peptide Vaccine

Glycoprotein D, administered intraperitoneally, protects mice against lethal challenge with HSV-1 and HSV-2 (PAOLETTI et al. 1984; CHAN 1983; LASKY et al. 1984; LONG et al. 1984). In addition, it has been shown that immunization with synthetic HSV-gD peptides can also confer protection in mice against a lethal HSV-2 challenge infection (EISENBERG et al. 1985). These protection experiments with synthetic peptides were carried out with the intention of inducing high virus-neutralizing antibody titers. Thus, the peptides were coupled to KLH, and animals were immunized i.p. multiple times (a total of four doses) and then challenged with HSV-2 by the same i.p. route. Animals which were immunized with the longer peptides 1–23(H), 1–23(1), and 1–23(2) were only partially (60%–90%) protected, and, furthermore, immunization with these peptides gave highly variable results. However, when the distal footpad route of immunization and viral challenge was used, the protective ability of these peptides was much higher and, in the case of the 1–23(1) peptide, all the immunized animals were protected. Sera from these immunized mice exhibited type-common neutralizing antibody, although the level of this activity did not seem to correlate with the degree of protection conferred (EISENBERG et al. 1985). These experiments clearly indicate that KLH-coupled HSV-gD peptides exhibit protective activity. However, vaccination with these peptides requires multiple injections of the KLH conjugates, and, therefore, the usefulness of such a synthetic vaccine is questionable.

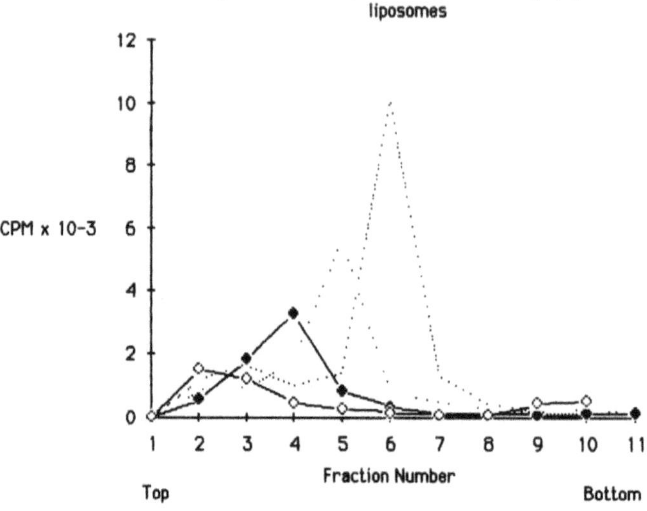

Fig. 7. ^{14}C-labeled palmitic acylated peptide-liposomes were layered on a continuous 5%–15% sucrose gradient and centrifuged to 35000 rpm in a Beckmann SW 50.1 rotor at 10 °C. Fractions were collected and counted for radioactivity, and density was determined for the peak fractions: ..., a density of 1.020; . . ., a density of 1.017; ♦–♦, a density of 1.014, and ◇–◇, a density of 1.010

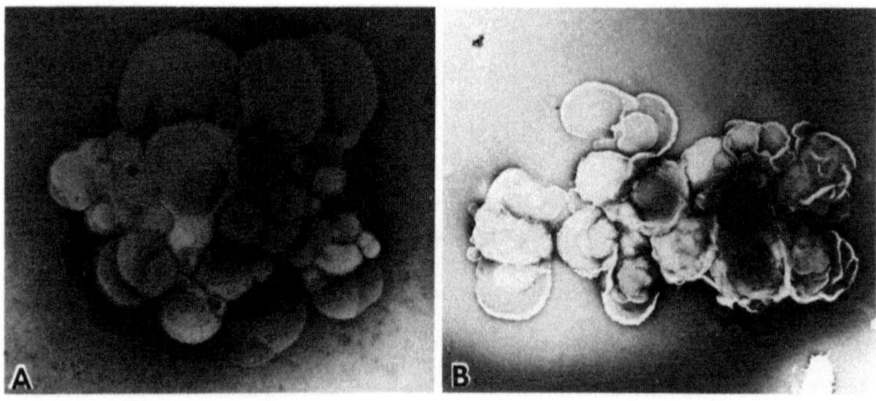

Fig. 8A, B. Electron micrographs of liposomes (**A**) or peptide-liposomes (**B**) at a peptide: lipid ratio of 4:1

Recently, it has been shown that the protective ability of viral envelope proteins was significantly increased when these antigens were incorporated into liposomes (THIBODEAU et al. 1981). In order to facilitate the insertion of the synthetic HSV-gD peptides into liposomes, the amino-terminus of the 1–23(2) peptide was acylated with palmitic acid (HOPP 1984) and the acylated peptide was then incorporated into liposomes, employing the method described by THIBODEAU et al. (1981). As demonstrated in Fig. 7, the acylated peptide-liposomes

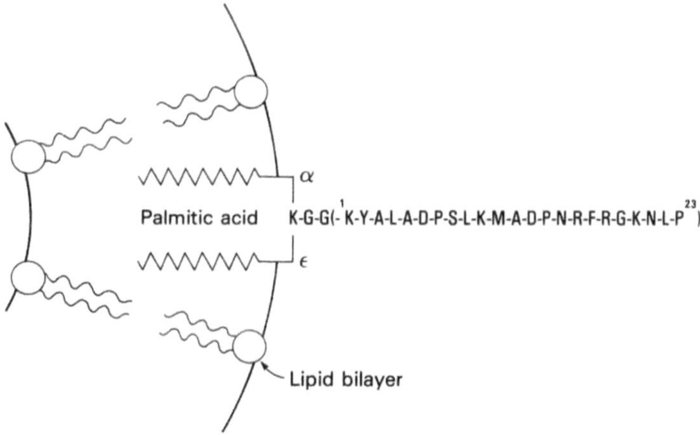

Fig. 9. The first 23 N-terminal amino acids of gD of HSV-2 are shown attached to a spacer, Gly Gly Lys. Two palmitic acid molecules were then coupled to the $\alpha+\varepsilon$ amino groups of lysin (K)

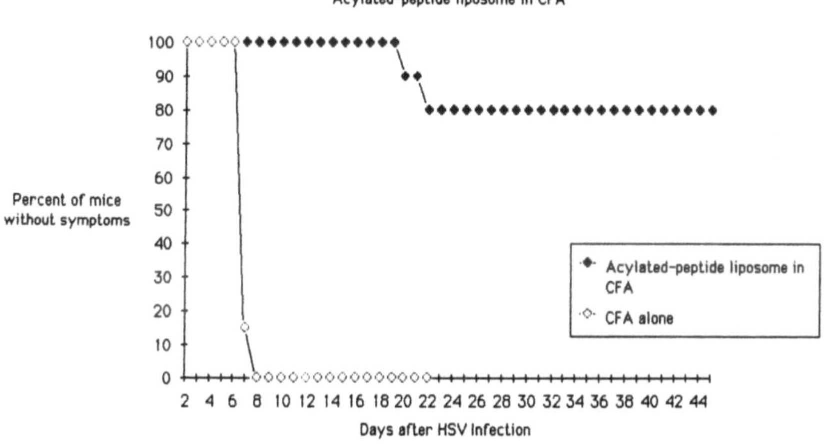

Fig. 10. Protection of Balb/c mice immunized with a single dose of acylated peptide-liposome and then challenged with a lethal dose of HSV-2. Immunization occurred intrafootpad 2.5 months before footpad challenge with a 7 LD 50 dose (2.2×10^6 PFU) of HSV-2 (WATARI et al. 1986)

are homogenous in regard to their buoyant density and, moreover, their buoyant density correlates with the amount of peptide incorporated. Electron microscopy (Fig. 8) indicates that the membrane of the peptide-liposomes exhibited a higher electron density compared to lipid membranes that do not contain peptide. This may indicate that the peptide, at least in part, is incorporated into the lipid membrane. The idealized structure of the peptide-liposome is depicted in Fig. 9.

Acylated peptide-liposomes were used to immunize mice against a lethal HSV-2 challenge. As demonstrated in Fig. 10, one dose of peptide-liposome injected in FCA 2.5 months before challenge protected 80% of the mice. This

Table 2. Anti-HSV antibody

Pooled serum from each group[a]	Neutralization titers for		Protection
	HSV-1	HSV-2	
FCA control	12	6	−
HSV-1 in FCA	389	97	+
Acylated peptide in FCA	6	8	−
Acylated peptide-liposome in FCA	5	5	+

[a] Balb/c mice were challenged with HSV-2 (10 LD 50) in the footpads 7 months after a single immunization of antigen in Freund's complete adjuvant. Bleedings were done 9 days after challenge with HSV-2

effect was long lived, as a single immunization of mice 7 months before challenge also resulted in significant protection (WATARI et al. 1986).

The most obvious explanation for this protective effect is the induction of a protective antibody response. We therefore determined the ability of sera from protected and unprotected mice to neutralize an infectious virus. To our surprise this in fact was not case (Table 2). Animals were immunized 7 months previously and then challenged with HSV-2 (a dose of 10 LD 50). It is clear that the antibody titers do not correlate with the protection seen. We consider the neutralizing antibody titer in the acylated peptide-liposome group to be baseline. We also found that antibody-binding data paralleled the neutralization data and supported the conclusion that protection occurs in the absence of an antibody response.

In considering mechanisms of protection other than antibody, we turned to the possibility of a viral-specific T cell. We had already shown that the 1–23(2) peptide could stimulate proliferating T cells in vitro. The presence of a T-cell response in these protected mice was determined by in vitro stimulation, from animals immunized with antigen 3 weeks previously. A typical T-cell response from mice primed to various forms of peptide, whether protective or not, can be seen in Table 3. These T cells could respond to peptide 1–23(2), gD, and HSV-infected cell lysates, and in fact responded equivalently. Thus, we could not distinguish the T-cell proliferative response in animals immunized with peptide in forms that were either protective or nonprotective. It is possible that different subpopulations of T cells make up this proliferating pool of cells. One interesting possibility is that we are inducing, besides proliferating cells, a population of suppressor T cells with the vaccine. Thus, we might predict that proliferating T cells which are immunopathic and enhance pathogenicity instead of protecting against it are induced by peptide alone or acyl peptide in the absence of liposome in FCA, or acylated peptide-liposome without FCA. This in fact appears to be the case. Another possibility is that cytotoxic T cells are induced by this acylated peptide-liposome.

It is important to note that the mixture of the three components, acylated peptide, liposome, and FCA, is essential for this protective response. One possible mechanism for this vaccine is that: (a) palmitic acid allows the incorporation of the peptide into the liposome; (b) the liposome can fuse with the membrane

Table 3. Activation of T cells by peptide and viral antigens

Antigen in culture	Dose	$\text{cpm} \times 10^{-3}$
1–23(2)	50 µg/ml	50.8
uv-HSV-1[a]	10^6 cells/ml	20.9
uv-HSV-2[a]	10^6 cells/ml	14.9
Purified gD	1 µg/ml	18.6
Ova	50 µg/ml	−1.1

[a] Infected cell lysate equivalents. Cell suspensions obtained from the popliteal and inguineal lymph nodes of animals immunized with acylated peptide-liposomes in FCA were passed over a nylon wool column, T cells purified, and then cultured with X-irradiated spleen (2500 R) plus antigen. These cultures were tested for responsiveness to antigen by cell proliferation measured by the incorporation of ^{3}H-thymidine after 3 days in culture

of an antigen-presenting cell; and (c) FCA acts as an adjuvant to induce an inflammatory response. In this way, one might imagine the induction of cytotoxic T cells by a peptide antigen. We are at present examining this possibility.

This method of immunization, i.e., protection in the absence of antibody, might have important consequences in latent infections. Thus, HSV one of the most common infectious agents of man, reactivates without a viremia but rather spreads between cells, and a cellular mechanism of protection may be required. There is clinical data supporting the notion that high anti-HSV-neutralizing antibody titers do not protect against reinfection, and, furthermore, there is evidence in man that antibody may play a negative role in protection against HSV.

References

Babbitt BP, Allen PH, Matseuda G, Haber E, Unanue ER (1985) The binding of antigen and the Ia molecule. Nature, 317:359

Balachandran N, Harnish D, Rawls WE, Bachetti S (1982) Glycoproteins of herpes simplex virus type 2 as defined by monoclonal antibodies. J Virol 44:344

Chan W (1983) Protective immunization of mice with specific HSV-1 glycoproteins. Immunology 49:343

Chou PW, Fasman GD (1978) The prediction of secondary structure of proteins from their amino acid sequence. Adv Enzymol 47:45

Cohen GH, Dietzschold B, Ponce de Leon M, Long D, Golub E, Varrichio A, Pereira L, Eisenberg RJ (1984) Localization and synthesis of an antigenic determinant of herpes simplex virus glycoprotein D that stimulates production of neutralizing antibody. J Virol 49:102

DeLisi C, Berzofsky JA (1985) T cell antigenic sites tend to be amphipathic structure. Proc Natl Acad Sci USA 82:7048

Dietzschold B, Eisenberg RJ, Ponce de Leon M, Golub E, Hudecz A, Varrichio A, Cohen GH (1984) Fine structure analysis of type-specific and type-common antigenic sites of herpes simplex virus glycoprotein D. J Virol 52:431

Eisenberg RJ, Ponce de Leon M, Cohen GH (1980) Comparative structural analysis of glycoprotein gD of herpes simplex virus types 1 and 2. J Virol 35:428

Eisenberg RJ, Long D, Pereira L, Hampar B, Zweig M, Cohen GH (1982a) Effect of monoclonal antibodies on limited proteolysis of native glycoprotein gD of herpes simplex virus type 1. J Virol 41:478

Eisenberg RJ, Ponce de Leon M, Pereira L, Long D, Cohen GH (1982b) Purification of glycoprotein gD of herpes simplex virus types 1 and 2 by use of monoclonal antibodies. J Virol 41:1099

Eisenberg R, Cerini CP, Heilman CJ, Joseph AD, Dietzschold B, Golub E, Long D, Ponce de Leon M, Cohen GH (1985) Synthetic glycoprotein D related peptides protect mice against herpes simplex challenge. J Virol 56:1014

Erickson BW, Merrifield RB (1976) Solid-phase peptide synthesis. In: Neurath H, Hill RL (eds) the proteins, vol 2. Academic, New York, p 255

Greenfield N, Fasman GD (1969) Computed circular dichroism spectra for the evaluation of protein conformation. Biochemistry 8:4108

Heber-Katz E, Schwartz RH, Matis L, Hannum C, Appella E, Fairwell T, Hansburg D (1982) The contribution of antigen-presenting cell I a to the specificity of T cell activation. JEM 155:1086

Heber-Katz E, Hansburg D, Schwartz RH (1983) The Ia molecule of the antigen-presenting cell plays a critical role in immune response gene regulation of T cell activation. JMCI 1:3

Heber-Katz E, Hollosi M, Dietzschold B, Hudecz F, Fasman GD (1985) The T cell response to the glycoprotein D of the herpes simplex virus: significance of antigen conformation. J Immunol 135:1385

Hopp TP (1984) Immunogenicity of a synthetic HBsAg peptide: enhancement by conjugation to a fatty acid carrier. Mol Immunol 21:13

Lasky LA, Dowbenko D (1984) DNA sequence analysis of type-common glycoprotein D genes of herpes simplex virus types 1 and 2. DNA 3:23

Lasky LA, Dowbenko D, Simonsen CC, Berman PW (1984) Protection of mice from lethal herpes simplex virus infection by vaccination with a secreted form of cloned glycoprotein D. Biotechnology 2:527

Long D, Madeira TJ, Ponce de Leon M, Cohen GH, Montgomery PC, Eisenberg RJ (1984) Glycoprotein D protects mice against a lethal challenge with herpes simplex virus types 1 and 2. Infect Immun 37:761

Paoletti DE, Lipinskas BR, Samsonoff C, Mercer S, Panicali (1984) Construction of live vaccines using genetically engineered poxviruses: biological activity of vaccinia virus recombinants expressing hepatitis B virus surface antigen and the herpes simplex virus glycoprotein D. Proc Natl Acad Sci USA 81:193

Pincus M, Gerewitz F, Schwartz RH, Scheraga H (1983) Correlation between the conformation of cytochrome c peptides and their stimulatory activity in a T lymphocyte proliferation assay. Proc Natl Acad Sci USA 80:3297

Showalter SD, Zweig M, Hampar B (1981) Monoclonal antibodies to herpes simplex virus type 1 protein, including the immediate-early protein ICP4. Infect Immun 34:684

Spear PG (1984) Glycoproteins specified by herpes simplex viruses. In: Roizman B (ed) Herpes viruses, vol. 3. Plenum, New York, p 315

Thibodeau L, Naud F, Baudreault A (1981) An influenza immunosome: its structure and antigenic properties. A model for a new type of vaccine. In: Nayak DP, Fox CF (eds) Genetic variation among influenza viruses. Academic, New York, p 587

Watari E, Dietzschold B, Heber-Katz E (1986) A synthetic peptide vaccine induces long-term protection in mice from a lethal HSV infection (to be published)

Watson RJ (1983) DNA sequence of the herpes simplex virus type 2 glycoprotein D gene. Gene 26:307

Watson RJ, Weis JH, Salstrom JS, Enquist LW (1982) Herpes simplex type I glycoprotein D gene: nucleotide sequence and expression in *E. coli*. Science 218:381

Antigen Presentation by B Lymphocytes:
A Critical Step in T–B Collaboration

A. Lanzavecchia

1   The Problem of the Antigen Bridge

In the very controversial field of T–B collaboration, one of the least controversial points is that specific T- and B-lymphocytes interact in the presence of antigen and that, following such interaction, B-lymphocytes proliferate and differentiate to produce antibody.

The way in which antigen mediates this interaction was investigated by Mitchison and his colleagues. Using T cells primed to one determinant (the carrier) and B cells primed to a different determinant (the hapten) they showed that effective T–B interaction only occurred if hapten and carrier were physically linked on the same molecule (Mitchison 1971). This requirement for physical linkage implied that the interaction between T and B cells involved physical contact between the two cells and that this contact was mediated by an antigen "bridge." The antigen, bound on one side by a B cell, was simultaneously recognized by a T cell and in this way T–B cell collaboration was triggered.

This interpretation is clouded by the fact that B cells and T cells recognize antigens in quite different ways (reviewed by Benacerraf 1978). B cells are capable of binding antigens directly through their surface Ig receptors, and the antibodies they produce will, in general, distinguish between the native and the denatured form of a given antigen. T cells, obtained by immunizing mice with a native protein antigen do not bind the antigen, and unlike B cells

The Basel Institute for Immunology, Grenzacher-Strasse 487, CH-4005 Basel

The Basel Institute for Immunology was founded and is supported by F. Hoffmann-La Roche, Ltd. Basel, Switzerland they do not distinguish between its native and denatured forms (CHESTNUT et al. 1980). The reason for this difference is that T helper cells can recognize foreign antigens only on the surface of specialized cells (therefore called antigen presenting cells, APC) and only in association with class II products of the major histocompatibility complex (MHC) (ROSENTHAL and SCHEVACH 1973; KATZ et al. 1973). Furthermore, in most cases foreign antigens which have been picked up by APC need to be degraded by the APC itself (a step called processing) before they can be recognized by T cells in association with class II molecules (UNANUE et al. 1984). Thus it seems that T-lymphocytes do not distinguish between native and denatured antigens simply because native antigens have to be degraded before they can be recognized by T cells.

Another important point, for which a theoretical analysis was provided (MATZINGER 1981) and which has recently gained strong experimental support (KAPPLER et al. 1981; DEMBIĆ et al. 1986), is that T cells have only one single receptor accounting for both antigen and MHC specificity. With a single receptor it is difficult to visualize how T-lymphocytes which have been activated against the bimolecular complex "processed antigen plus MHC" on an APC can use the same specificity to bind to the trimolecular complex "native antigen bound to surface immunoglobulins plus MHC" on the surface of a specific B cell. Recent evidence suggests that the simple concept of the antigen bridge is not as simple as it first appeared.

2 B-Lymphocytes as Antigen Presenting Cells

An alternative to the antigen bridge is the suggestion that B cells do not present antigen bound to surface immunoglobulins (SIg) but rather that they internalize the antigen and process it in the same way as conventional APCs and finally present this processed antigen to T cells in association with class II molecules. The model postulates that SIg serve as specific receptors for antigen and enable specific B-lymphocytes to selectively concentrate antigens from the solution.

Evidence for the role of SIg in concentrating antigen for presentation to T cells has been provided by ROCK et al. (1984), who showed that mouse hapten-specific B-lymphocytes are very efficient in presenting hapten-carrier conjugates to carrier-specific T-cell hybridomas. Evidence that B-lymphocytes are capable of processing soluble antigens comes from experiments in both the mouse and human systems using B-cell tumors (CHESTNUT et al. 1982), activated B cells (CHESTNUT and GREY 1981), or B cells transformed by Epstein–Barr virus (EBV) (CHU et al. 1984). In addition, it has been shown that mouse B-lymphocytes will present rabbit-anti mouse Ig to rabbit-Ig-specific T cells much more efficiently than they will present normal rabbit Ig, suggesting again a role of antigen concentration by SIg (CHESTNUT and GREY 1981).

In the past, it has been extremely difficult to study the antigen bridge using conventional antigens because specific T and B cells occur at such extremely low frequencies. For instance, the frequency of B cells specific for tetanus toxoid (TT) in the peripheral blood of immunized humans is in the order of one in 10^3–10^4 B cells and in nonimmune donors such cells are practically undetectable (LANZAVECCHIA et al. 1983). To overcome this problem we have developed a clonal system to study T–B interaction using human T-cell clones and EBV-transformed B-cell clones specific for TT.

3 Specific T- and B-Lymphocytes Interact in the Presence of Antigen

Several human T-cell clones specific for TT were isolated and maintained in long term culture by a standard method using allogeneic peripheral blood mononuclear cells (PBM) and phytohemagglutinin (PHA) for restimulation. The use of allogeneic PBM ensures that T cells are not contaminated by any autologous APC. In addition, the T cells were used in all the experiments after expansions for 3–4 weeks in IL-2 alone, a procedure which results in a pure T-cell population devoid of any APC. Three independent clones of EBV-transformed B cells producing specific anti-TT IgG_1 antibodies were isolated by limiting dilution (LANZAVECCHIA 1985). Having pure T- and B-cell clones specific for the same antigen, it was possible to assess their interaction in the presence of antigen. There are several ways in which such an interaction could be measured and three of them are shown in Fig. 1.

When irradiated B cells are mixed with T-cell clones, a T-cell proliferative response can be observed in the presence of antigen. This assay measures presentation of TT by B to T cells resulting in T-cell proliferation. Figure 1 also shows that when B cells are mixed with irradiated T cells, the B-cell proliferation (B cells are EBV-transformed) is dramatically reduced in the presence of TT.

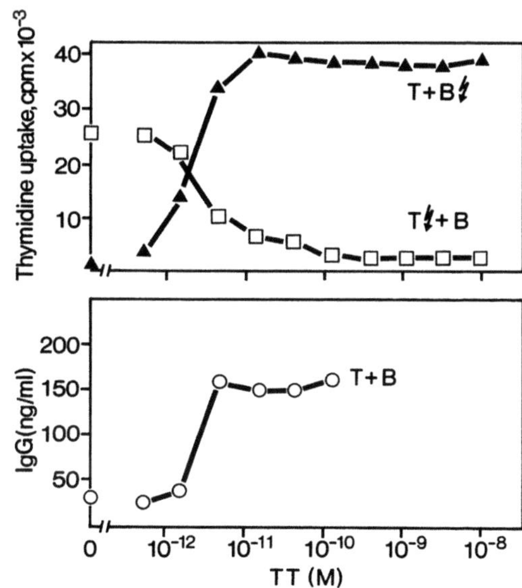

Fig. 1. Antigen mediated interaction between a TT-specific T-cell clone and a TT-specific EBV-transformed B-cell clone: 2×10^4 T cells from a TT-specific T-cell clone (KT-4) were cultured in 200 µl RPMI-10% FCS in flat bottomed microplates with 2×10^4 B cells from an EBV-transformed B-cell clone (4.2) in the presence of various concentrations of TT. In the *upper panel* either T cells or B cells were irradiated (3000 R) and thymidine uptake by co-cultured B (□) or T (▲) cells was measured on day 2. In the *lower panel* the level of IgG in the culture supernatant was measured after 48 h with an enzyme immunoassay

^{51}Cr-release assays as well as immunofluorescent staining of the cells in culture demonstrate that such decreased proliferation of EBV-transformed B cells is not due to a cytotoxic effect, but rather to the induction of terminal differentiation. This was documented by a marked increase (from 10% to 50%) in cells containing intracytoplasmic Ig and an increase in antibody production measured in the culture supernatant. In addition IL-2 and γ-IFN can be measured in the supernatant of T and B cells cultured in the presence of antigen (data not shown). Remarkably, all the above effects have a similar antigen dose dependence. These findings suggest that a whole series of cooperative effects, including lymphokine release and induction of B-cell differentiation, can be triggered by the direct interaction between T-cell clones and EBV-transformed B cells. The antigen presentation assay was finally chosen as a measure of the T–B interaction since T-cell triggering is required for the delivery of help to B cells and the triggering of T-cell proliferation can be easily mesured.

4   Antigen Uptake by Surface Immunoglobulins

The uptake of antigen by APC is obviously an important step in antigen presentation. It is therefore conceivable that antigen-specific B cells would be more efficient in presenting that antigen to T cells because of their ability to bind the antigen to SIg. This was tested in a series of experiments by comparing the efficiency of TT-specific EBV-transformed B cells, TT-nonspecific EBV-transformed B cells, and PBM in presenting TT to TT-specific T-cell clones.

Table 1 summarizes the data of several experiments performed over a period of several months. Although we could observe some variability in the responsiveness to TT of some T-cell clones (i.e., shifts in the antigen dose response curve related to the time elapsed after the last antigenic restimulation), the differences in the dose of antigen required with the different APC for the induction of 30% maximum response were consistent from experiment to experiment and from clone to clone. The overall picture that emerges indicates a clear hierarchy in the capacity to present antigen to T cells:

1. Specific B cells are always much more efficient in presenting antigen than nonspecific B cells or PBM, but the difference may vary from 10^2 to 10^4 times or more according to the T-cell clone tested.
2. There is some preferential pairing between specific T and B cells in that some T-cell clones perform better with one specific B-cell clone than with another.

Four series of experiments show that the high efficiency of TT presentation by TT-specific B cells is dependent on the specificity of the SIg, which allows more effective antigen uptake (LANZAVECCHIA 1985):

1. Denatured, reduced and carboxymethylated TT (RCM–TT) has lost the capacity to react with the antibodies produced by the EBV-transformed B cells but still fully retains the capacity to stimulate T cells. When RCM-TT

Table 1. Concentration of TT (ng/ml) required for induction of 30% maximum T-cell stimulation using different combinations of T-cell clones, TT-specific B cells, TT-nonspecific B cells and PBM.

TT-specific T-cell clone	Antigen presenting cells				
	TT-specific			TT-non-specific	
	8.5	11.3	4.2	7.1	PBM
KT-30	0.01	0.05	0.1	20×10^3	20×10^3
KT-42	0.07	0.07	nd	2×10^3	1×10^3
KT-12	0.07	5	1	20×10^3	20×10^3
KT-4	0.2	0.2	5	1×10^3	1×10^3
KT-26	1	5	20	10×10^3	10×10^3
KT-2	10	5	0.5	5×10^3	3×10^3
KT-1	10	20	3	4×10^3	1×10^3
KT-18	20	20	nd	20×10^3	20×10^3
KT-31	20	20	nd	5×10^3	10×10^3
KT-20	70	70	nd	20×10^3	5×10^3
KT-32	100	70	200	20×10^3	4×10^3

In total 2×10^4 T cells were cultured with 2×10^4 3000 R-irradiated cells from TT-specific B-cell clones (8.5, 11.3, and 4.2), or from a TT-nonspecific B-cell clone (7.1) or with 10^5 irradiated autologous PBM as APC. For each combination TT was titrated and the approximate concentration (ng/ml required to induce 30% of the plateau level of proliferation (obtained with a specific B cell as APC) was determined. Data represent the average mean of five different experiments. nd, not determined

was used, no significant difference was found between specific and nonspecific B cells in the efficiency of presentation.

2. No differences were observed when TT-specific and TT-nonspecific B cells were tested for their capacity to present an unrelated antigen (PPD).
3. Anti-Ig antibodies interfere with antigen presentation by blocking uptake of antigen and not by interfering at later stages.
4. Antibodies against native antigen block antigen uptake by specific B cells in an epitope specific way.

The last point, which illustrates the potential regulatory role of antibodies, is shown in Fig. 2. Monoclonal anti-TT antibodies were purified from the culture supernatant of EBV-transformed B-cell clones and tested for their effect in the antigen presentation assay. When PBM were used as APC, the addition of anti-TT resulted in an increased efficiency of presentation, as shown by a 10–20 times reduction in the concentration of antigen required for presentation to T cells. The opposite effect was observed when TT-specific B cells were used as APC. In this case the addition of anti-TT antibody (but not of an irrelevant antibody) resulted in a marked decrease in the efficiency of antigen presentation. This is probably due to competition for antigen binding between soluble and cell-bound antibodies. Furthermore, such competition by soluble antibodies was more evident in the homologous than in the heterologous combination.

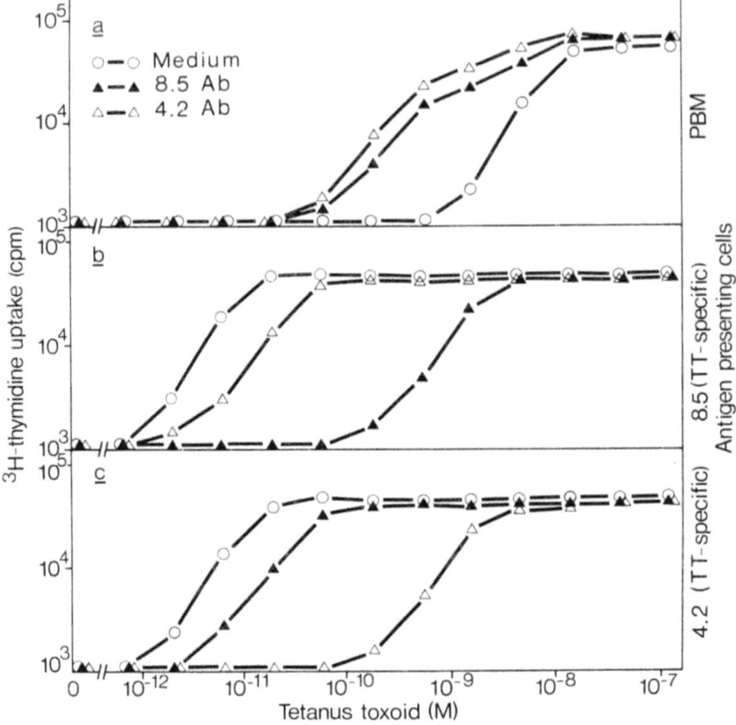

Fig. 2 a–c. The effect of antibody aginst the antigen on antigen presentation by PBM or specific B cells: 2×10^4 T cells from a TT-specific T-cell clone (KT 4) were cultured with either 10^5 irradiated PBM (**a**) or with 2×10^4 irradiated TT-specific EBV-transformed B cells 4.2 or 8.5 (**b, c**) in the presence of different concentrations of TT. Monoclonal anti-TT antibodies (from culture supernatants) were added at 10 µg/ml at the initiation of cultures. Thymidine uptake was measured on day 2

5 Evidence that Antigen is Processed by Specific B-Lymphocytes

The crucial question about the antigen bridge is whether specific B cells present the antigen to T cells bound to SIg or whether intracellular antigen processing is required for T–B intraction. This question was explored by separating binding of antigen to B cells from presentation to T cells. TT-specific B cells were first pulsed with low concentrations of TT in the cold to allow antigen binding, then washed and incubated at 37° C for different periods of time before being fixed with glutaraldehyde. As is evident from Table 2, B cells that have been pulsed with antigen, washed, and fixed immediately were not capable of triggering T-cell proliferation, although they did bind TT. However, the same cells were able to trigger T-cell proliferation if, after pulsing, they were incubated for 30–60 min at 37° C before they were fixed. If such incubation was performed in the presence of 10^{-5} M chloroquine (which inhibits antigen processing by increasing the pH of endosomic vesicles) (ZIEGLER and UNANUE 1982) or of

Table 2. Evidence for antigen processing by specific B cells.

Length of incubation at 37° C before fixation (min)	^{3}H-thymidine uptake (cpm) by T cells	
	KT-4	KT-30
0	161	90
10	280	207
30	7309	4985
90	65273	71050
90 (chloroquine)	1255	900
90 (azide and deoxyglucose)	2022	2955

TT-specific B cells (clone 11.3) were incubated with $10^{-10}\,M$ TT at 4° C for 30 min, washed, and incubated at 37° C in RPMI-FCS for different lengths of time before being fixed with glutaraldehyde (LANZAVECCHIA 1985). The incubation was performed in medium alone or in medium containing chloroquine ($10^{-5}\,M$). Fixed B cells (2×10^4) were cultured with T cells (2×10^4) in 200 µl RPMI-FCS in U-bottomed microplates. Thymidine uptake was measured on day 2

azide and deoxyglucose (which inhibit cellular metabolism), then the B cells did not acquire the capacity to trigger T cells.

Additional experiments showed that trypsin treatment of B cells immediately after pulsing with antigen completely abolished antigen presentation. The same treatment, however, had no effect on B cells that had been incubated for 1 h at 37° C after the antigen pulse (data not shown). These results suggest that TT is first present bound to SIg (which are sensitive to trypsin) but, following incubation at 37° C, fragments of TT are internalized or present on the cell surface in a trypsin insensitive form, most likely associated with class II molecules.

Taken as a whole, these data demonstrate that, in the case of a complex protein antigen like TT, B cells use SIg to concentrate the antigen, but not to present it to T cells. In fact antigen has to be processed intracellularly before it can be displayed on the cell surface for recognition by T cells.

The nature of antigen processing is not at all clear and at the moment we are left only with hypotheses. The simplest one which has some experimental basis is that, as T cells must recognize antigen and MHC with one single receptor, the size of the antigen must be suitable to allow such trimolecular interactions and, in addition, the antigen might need some as yet poorly defined physiochemical properties that facilitate binding to MHC molecules. Large proteins are unlikely to be suitable for this type of interaction, but small peptides are good candidates. That the size of the antigen is crucial for T-cell recognition is a well-known fact: small haptens and peptides do not require processing, while large protein antigens do. In addition to simply reducing the antigen size, processing might reveal buried peptides which have appropriate characteristics of hydrophobicity (or amphipathicity) (DELISI and BERZOWSKI 1985) to become associated with the cell membrane or with class II molecules. Experimental evidence for both hypotheses has recently been produced. SHIMONKEVITZ et al. (1983) have shown that fixed APC, which are themselves not able to process and present native antigens to T cells, can present peptides of antigen obtained by in vitro digestion with proteolytic enzymes. This technique has some potential applicability in mapping the epitope specificity of T cells.

6 Presentation to T Cells

T–B collaboration has been shown to be MHC restricted (KATZ et al. 1973). Therefore the finding that antigen presentation of TT by TT-specific B-lymphocytes is also MHC restricted is certainly not unexpected. This is shown by the fact that: (a) TT-specific B cells present TT to autologous but not to allogeneic TT-specific T-cell clones, and (b) anti-class II antibodies inhibit antigen presentation to T cells. If the structure recognized by T cells is processed antigen in association with class II molecules, it is likely that inhibition by anti-class II antibodies (resulting in a decrease in available class II molecules) could be reversed by increasing the concentration of antigen. This pattern of competitive inhibition is exactly what is observed in Fig. 3, where anti-class II antibodies inhibit antigen presentation only at low and not a high concentrations of antigen.

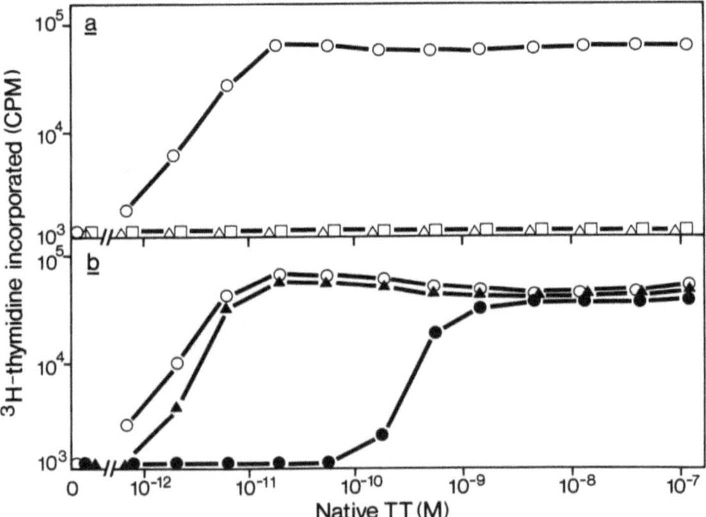

Fig. 3 a, b. Antigen presentation by antigen-specific B cells is MHC-restricted. **a** A TT-specific B-cell clone (11.3) was irradiated and cultured with an autologous TT-specific T-cell clone (o), or with allogeneic TT-specific clones (Δ) and (□). **b** TT-specific T and B cells (clones KT4 and 11.3) were stimulated with TT in the presence of anti-HLA-DR monoclonal antibody D1.12 at 10 µg/ml (●), anti-HLA-A, B, C antibody W6/32 at 50 µg/ml (▲), or medium alone (o). Reprinted by permission from Nature 314:537

Once antigen has been taken up and processed by B cells, antigen presentation can no longer be inhibited by anti-Ig antibodies, or by antibodies against the antigen (UHR and MÖLLER 1968). This is consistent with the notion that SIg play a role only in antigen uptake and that the immunogenic moiety recognized by T cells consists of processed antigen, which in most instances does not react with antibodies directed against the native antigen. There are two exceptions. Reports by CORRADIN and ENGERS (1984) and by LAMB et al. (1984) show that antibodies directed against a peptide recognized by T-cell clones were able to block presentation of this peptide to T cells, although such blocking was fairly inefficient. This might be expected considering that: (a) immunogenic determinants may be very scanty on APC; (b) although T-cell specificity was mapped using these artificially produced peptides there is no evidence that the APC will produce exactly the same peptide or rather overlapping peptides; and (c) the conformation of the peptides may be influenced by their interaction with class II molecules, therefore antibodies raised against the peptide on a carrier molecule might have problems recognizing the same peptide associated with class II molecules. In general, immunization with either native or denatured antigens does not lead to the production of antibodies with the capacity to block T-cell antigen recognition (CHESTNUT et al. 1980). The effect of antibodies is therefore limited to the phase of antigen uptake. When antibodies are surface bound, as they are in B cells, they allow very efficient antigen uptake and, even in solution, they may increase antigen uptake by Fc-receptor positive APC. Soluble antibodies against the antigen might also have an inhibitory effect on the antibody response since they can effectively compete with antigen uptake by specific B cells (see above), an effect which is likely to contribute to the spread of the antibody response from one epitope to another of the same molecule. It is easy to predict that there will be more information on the elusive antigenic moiety recognized by T cells when molecularly engineered soluble T-cell receptors will be available for binding studies with antigen-pulsed APC.

7 The Paradox of the High Affinity of T–B Interaction

In the experiments described specific B and T cells are able to interact with each other in the presence of 10^{-13}–10^{-11} M TT. This antigen concentration is surprisingly low if we consider that only the B cells are capable of binding antigen and that the affinity of surface antibodies for TT is in the order of 10^8 $1\,\text{mol}^{-1}$. It follows that only a very small fraction of the SIg needs to be occupied by antigen at any given time in order to allow B cells to interact with T cells. These findings seriously challenge the concept that extensive cross-linking of SIg by antigen (which requires nearly saturating concentrations of a bivalent ligand) is an essential signal in B-cell activation. Although cross-linking of SIg by anti-Ig antibodies is known to lead to a series of events which culminate in increased class II expression, membrane depolarization, and induction of responsiveness to growth factors, it might well be that cross-linking of SIg is not required when B cells are activated by interacting with a T helper cell. A strong argument against the requirement for SIg cross-linking is that soluble protein antigens generally carry epitopes which are monovalent for a single B cell and therefore cannot cross-link SIg receptors. Evidence against a requirement for cross-linking has been reported recently by TONY et al. (1985) who showed that monovalent rabbit anti-mouse Ig is as efficient as bivalent antibodies in triggering a polyclonal activation of mouse B cells by rabbit Ig-specific T helper cells.

The extremely high affinity of T–B intraction becomes perhaps less surprising if we consider that nonspecific B-lymphocytes or PBM can interact with TT-specific T cells in the presence of 10^{-7}–10^{-8} M TT: conditions in which neither B cells nor PBM nor T cells have any measurable binding affinity for the free antigen.

Three considerations might help to explain the paradox of the high affinity with which T cells interact with B cells or APC, compared to the much lower affinity with which B cells or APC bind to the antigen. The first is that the continuous internalization of antigen (HOWARD 1985) followed by intracellular processing may lead to accumulation of processed determinants in the APC, thus allowing the display on the cell surface of a far greater number of determinants than the number of native antigen molecules bound at a given time. This model predicts that in general only peptides which accumulate in the APC will be immunogenic for T cells. An understanding of the basis for such selective accumulation would help in elucidating the properties of the antigenic determinants recognized by T cells and perhaps would allow the modification of natural peptides so as to increase their interaction with APC.

The second possibility is that the immune system has developed receptor-ligand systems designed to facilitate cellular interactions in a nonspecific way. One of these systems may involve the T4 antigen present on class II-restricted T cells, which is supposed to interact with the constant part of class II molecules on APC (MARRACK et al. 1983). When anti-T4 monoclonal antibodies are added to specific T and B cells in the presence of antigen we observe a shift in the dose-response curve towards higher concentrations of antigen (LANZAVECCHIA and BOVE 1985). This suggests that the role of T4 is to decrease the concentration of antigen required for T-cell triggering, perhaps by allowing more extensive membrane contact. However, such a mechanism does not affect the specificity of the T-cell response, since ligation of T4 antigen will not deliver a triggering signal to T cells.

Finally it must be stressed that it is not known how many epitopes must be present on an APC in order to trigger T-cell responses. This number will of course depend on the intrinsic affinity of the T-cell receptor, but it can be anticipated that in some instances it will be very low. A maximum estimate can be made in the case of clone KT30 (TT specific) proliferating in response to 10^{-13} M TT presented by a TT-specific B-cell clone. In these conditions it can be calculated that not more than 1000 molecules are available in solution for each B cell; therefore the number of determinants made available for T cells must necessarily be only a fraction of this number. The definitive answer to this question will come when the T-cell receptor, which is the only ligand useful for such studies, will be available in a cell-free form.

8 What Can this New View of the Bridge Contribute to New Strategies of Vaccination?

The goal of vaccination is to induce an appropriately protective immune response without inducing the disease itself or other undesirable effects. In the past this was achieved by giving inactivated, attenuated, or partially purified pathogens with all the concurrent side effects. A currently pursued alternative is the identification and isolation (or construction) of antigenic determinants that are recognized by protective antibodies and the immunization with such isolated determinants (either in the form of a peptide or of an anti-idiotypic antibody that bears the internal image of the antigen) to stimulate a protective antibody response. These modern methods are not themselves problem free for the following reasons:

1. Neither artificially constructed determinants, nor internal image antibodies are a faithful image of the pathogen; therefore the antibodies raised against them will bind the pathogen only by cross-reaction.
2. Peptides are not themselves immunogenic for B cells, since they usually do not contain determinants recognized by both T and B cells and therefore have to be coupled to a carrier protein.

Although such carriers can certainly be made, the problem comes when the immune system primed in this way is challenged with the natural infectious agent. This is equivalent to giving the right hapten on the wrong carrier, i.e., it may not result in a secondary antibody response (MITCHISON 1971). It follows that such an approach would be useful only in cases where a secondary response is not required. Although the construction of special carriers is unnecessary where the vaccination is done with internal image antibodies (carrier determinants exist on both the constant and variable region, JORGENSEN et al. 1983) the resulting response will also be against an unfaithful image attached to the wrong carrier.

One way to avoid these problems is to reconstruct a functional hapten-carrier conjugate by isolating from the same infectious agent a peptide that can be recognized by T helper cells and coupling this to the B-cell determinant. Although this approach seems quite cumbersome, the techniques to identify and construct T- and B-cell epitopes are already available. One possible disastrous effect of "tailoring" antigens is the suppression of the response to other antigenic determinants of the same molecule, known as "original antigenic sin." For example, vaccination with a construction which was missing the truly protective determinant found on the natural pathogen could permanently impair the individual's ability to subsequently mount a protective response against that pathogen.

Although the mechanisms of such epitope-specific suppression are unknown, it seems a likely possibility that circulating antibodies against the carrier could inhibit antigen uptake by unprimed B cells responding against the new determinant.

In some instances it might be favorable to prime T cells without obtaining an antibody response against the antigen. This can be achieved by immunizing with denatured or fragmented antigen. In this case any antibody that is made will react with the denatured antigen and not with the native antigen.

In principle, the antibody response against some antigens like polysaccharides may be difficult to obtain because of the lack of appropriate carrier determinants. The coupling to such antigens of a determinant that can be recognized by T helper cells could be expected to increase the antibody response and to expand specific B cells.

9 Open Questions

Can antigen-specific resting B cells present antigen in such a way as to prime resting T cells? The experimental answer to this question is certainly a problem (a) because of the difficulties in defining "resting" (or "virgin" or "primary") T and B cells, and (b) because the frequency of specific resting T and B cells is too low to allow in vitro experiments. A possible answer to this question may come from a chicken in vivo system in which B cells and macrophages are of different MHC types (LASSILA and MATZINGER, personal communication). There is some accumulating evidence that resting B cells may be able to trigger primed T cells (ASHWELL et al. 1984).

If B cells can present antigen to T cells, what is the role of macrophages and other APC in the induction of an antibody response? Are macrophages required in a primary response only because specific B cells are too rare or do they exert some additional functions that B cells do not? It is possible that macrophages are required because B cells are not able to trigger a resting T cell. A reason for this might stem from the requirements for self tolerance. Tolerance to self antigens has a given affinity threshold dependent on the APC in the thymus, which must be the same as the triggering threshold for T cells in order to maintain tolerance to self and response to foreign antigens (BRETCHER and COHN 1970). It would be therefore extremely dangerous if autoreactive B cells (which can arise for instance by somatic mutation) would be able to present antigen to prime virgin T cells. T cells which are tolerant to autoantigens presented by thymic APC may not be tolerant to the same antigen presented by a specific B cell, which is 10^3–10^4 times more efficient. Therefore both tolerance induction and priming must remain the domain of the same APC.

Since antigen is recognized sequentially by B and T cells one would expect no constraints as to the epitope specificity of interacting T and B cells. There might be, however, exceptions to this rule. It is known that antibodies can exert profound effects on antigen by stabilizing or destroying its conformation (CELADA et al. 1983) or by protecting it from denaturation (MELCHERS and MESSER 1970). Therefore it is possible that in a B cell the specificity of the antibody might influence the way in which antigen is processed and would therefore result in the display on the surface of a spectrum of peptides slightly different from the one displayed by other B cells or by conventional APC. This hypothesis which has some experimental support (MANCA et al. 1985; see also Table 1) remains to be tested.

Last but not least is the question of how T cells help B cells to divide and make antibodies. We have so far considered only the first step of T–B interaction, namely the antigen bridge; although there are other ways of activating B cells (such as mitogens or alloreactive T cells) the bridge is responsible for the characteristic which is the hallmark of the immune system: the antigen specificity.

References

Ashwell JD, DeFranco AL, Paul WE, Schwartz RH (1984) Antigen presentation by resting B cells; radiosensitivity of the antigen presentation function and two distinct pathways of T cell activation. J Exp Med 159:881

Benacerraf B (1978) A hypothesis to relate the specificity of T lymphocytes and the activity of I-region-specific IR genes in macrophages and B lymphocytes. J Immunol 120:1809–1811

Bretcher P, Cohn M (1970) A theory of self–nonself discrimination. Science. 169:1042

Celada F, Schumaker VN, Sercarz E (eds) (1983) Protein conformation as immunological signal. Plenum, New York

Chestnut R, Grey HM (1981) Studies on the capacity of B cells to serve as antigen presenting cells. J Immunol 126:1075–1079

Chestnut R, Endres R, Grey H (1980) Antigen recognition by T cells and B cells: recognition of cross-reactivity between native and denatured forms of globular antigens. Clin Immunol Immunopathol 15:397

Chestnut RW, Colon SM, Grey HM (1982) Requirement for the processing of antigens by antigen-presenting B cells. J Immunol 129:2382–2388

Chu E, Umetsu D, Lareau M, Schneeberger E, Geha RA (1984) Analysis of antigen uptake and presentation by Epstein–Barr virus-transformed human lymphoblastoid B cells. Eur J Immunol 14:291–298

Corradin G, Engers HD (1984) Inhibition of antigen induced T cell clone proliferation by antigen-specific antibodies. Nature 308:547–548

DeLisi C, Berzowski JA (1985) T cell antigenic sites tend to be amphypathic structures. Proc Natl Acad Sci USA 82:7048–7052

Dembić Z, Haas W, Weiss S, McCubrey J, Kiefer H, von Boehmer H, Steinmetz M (1986) Transfer of specificity by murine α and β T-cell receptor genes. Nature 320:232–238

Howard JC (1985) Immunological help at last. Nature 314:494–495

Jorgensen T, Bogen B, Hannestad K (1983) T helper cells recognize an idiotope located on peptide 88 ≠ 114/117 of the light chain variable domain of an isologous myeloma protein (315). J Exp Med 158:2183

Kappler JW, Skidmore B, White J, Marrack P (1981) Antigen inducible, H-2 restricted, interleukin-2-producing T cell hybridomas: lack of independent antigen and MHC-recognition. J Exp Med 153:1198

Katz DH, Hamaoka ME, Dorf ME, Benacerraf B (1973) Cell interactions between histoincompatible T and B lymphocytes. The H-2 gene complex determined successful physiologic lymphocyte interactions. Proc Natl Acad Sci USA 70:2624–2628

Lamb JR, Zanders ED, Lake P, Webster RG, Eckels DD, Woody JN, Green N, Lerver RA, Feldman M (1984) Inhibition of T cell proliferation by antibodies to synthetic peptides. Eur J Immunol 14:153–157

Lanzavecchia A (1985) Antigen-specific interaction between T and B cells. Nature 314:537–539

Lanzavecchia A, Bove S (1985) Specific B lymphocytes efficiently pick up, process and present antigen to T cells. Behring Inst Mitt 77:82–87

Lanzavecchia A, Parodi B, Celada F (1983) Activation of human B lymphocytes: frequency of antigen-specific B cells triggered by alloreactive or by antigen-specific T cell clones. Eur J Immunol 13:733–738

Manca F, Kunkl A, Fenoglio D, Fowler A, Sercaz E, Celada F (1985) Constraints in T–B cooperation related to epitope topology in *E. coli* B-galactosidase I. The fine specificity of T cells dictates the fine specificity of antibodies directed to conformation-dependent determinants. Eur J Immunol 15:345

Marrack P, Endres R, Shimonkevitz R, Zlotnik A, Dialynas D, Fitch F, Kappler J (1983) The MHC-restricted antigen receptor on T cells. II. Role of the L3T4 product. J Exp Med 158:1077–1091

Matzinger P (1981) A one receptor view of T cell behavior. Nature 292:497–502

Melchers F, Messer W (1970) Enhanced stability against heat denaturation of E. Coli wild type and mutant β-galactosidase in the presence of specific antibodies. Biochem Biophys Res Commun 40:570

Mitchison NA (1971) The carrier effect in the secondary response to hapten-protein conjugates. II. Cell cooperation. Eur J Immunol 1:18–27

Rock KL, Benacerraf B, Abbas AK (1984) Antigen presentation by hapten-specific B lymphocytes. J Exp Med 160:1102–1113

Rosenthal AS, Shevach EM (1973) Function of macrophages in antigen recognition by guinea pig T lymphocytes. I. Requirement for histocompatible macrophages and lymphocytes. J Exp Med 138:1194–1212

Shimonkevitz R, Kappler J, Marrack P, Grey HM (1983) Antigen recognition by H-2 restricted T cells. II. Cell free antigen-processing. J Exp Med 158:303

Tony HP, Phillips NE, Parker DC (1985) Role of membrane Ig crosslinking in membrane Ig-mediated major histocompatibility restricted T cell-B cell cooperation. J Exp Med 162:1695–1708

Uhr J, Möller G (1968) Regulatory effect of antibody on the immune response. Adv Immunol 8:81–126

Unanue ER, Beller DI, Lu CY, Allen PM (1984) Antigen presentation: comments on its regulation and mechanism. J Immunol 132:1–5

Ziegler HK, Unanue E (1982) Decrease in macrophage antigen catabolism caused by ammonia and chloroquine with inhibition of antigen presentation to T cells. Proc Natl Acad Sci USA 79:175–178

The Value of Synthetic Peptides as Vaccines for Eliciting T-Cell Immunity

R.H. SCHWARTZ

1 Introduction

As our understanding of T-cell activation has increased at the molecular level, several investigators have begun to discuss the idea of using synthetic peptides as vaccines for inducing T-cell immunity (LERNER 1984; LAVER and AIR 1985). My view as to the value of such an approach is rather pessimistic. In this paper I will briefly outline our current understanding of antigen recognition by T cells and then discuss in detail what I see to be the major pitfalls in the peptide approach to vaccine development. I do not mean to imply in this discussion that developing vaccines to induce T-cell immunity represents a fruitless endeavor. Rather, I wish to stress that the synthetic peptide approach may not be the most rational one to choose or the most cost-effective at the present time.

2 T Cell Activation

2.1 Helper Cells

Most antigen-specific helper or inducer T cells recognize foreign antigens in association with class II (Ia) molecules of the major histocompatibility complex (MHC) (SCHWARTZ 1984). Activation of the T cell requires the antigen-specific receptor to be simultaneously occupied by both the Ia molecule and the foreign antigen (SCHWARTZ 1985). This ternary complex may exist in the form of a

Laboratory of Immunology, National Institute of Allergy and Infectious Diseases, National Institutes of Health, Bethesda, MD 20892, USA

Current Topics in Microbiology and Immunology, Vol. 130

trimolecular structure, in which all three entities physically interact, or it may involve separate binding sites on the receptor(s) for the antigen and the Ia molecule. The bulk of the data at the present time favors the former model (SCHWARTZ 1985).

The form of the foreign antigen that is required for T-cell activation in many cases appears to be a peptide fragment of the original molecule. This has been demonstrated by several laboratories with aldehyde fixation and specific and nonspecific inhibitors of lysosomal enzymes to block antigen presentation by B cells and macrophages (UNANUE 1984). This blockade can be circumvented by using chemical or enzymatic fragments of the protein or in some cases by simply denaturing the molecule (SHIMONKEVITZ et al. 1983; ALLEN and UNANUE 1984; STREICHER et al. 1984; KOVAC and SCHWARTZ 1985). These studies suggested that processing involves only alteration of the antigen to expose critical residues for recognition by either the T-cell receptor or the Ia molecule.

Interaction between the antigenic peptide (or denatured molecule) and the Ia molecule, either prior to or during T-cell recognition, creates a problem for the immune system. Because certain antigenic determinants do not interact favorably with the Ia molecule(s) possessed by a given individual, T-cell clones specific for this particular combination of determinant and Ia molecule cannot be activated (SCHWARTZ 1984). In some cases the immune system is capable of recognizing a different determinant on the same antigen. However, in other cases, no determinants with the desired associative properties are generated or exposed. In the latter cases the individual is a nonresponder to the antigen. Although this may not be the only mechanism of *Ir* gene control for a helper T-cell response, it has been clearly demonstrated to play a critical role in several well-defined situations (SCHWARTZ 1985; BABBITT et al. 1985).

2.2 Cytotoxic T Cells

The role of cytotoxic T lymphocytes (CTL) in resisting viral infections has been suggested in a number of disease models (ZINKERNAGEL and DOHERTY 1979). Originally, it was felt that CTL recognized intact membrane glycoproteins in association with class I molecules of the MHC (GEIGER et al. 1979). This concept would imply that the antigen-specific receptor on CTL is fundamentally different from that on helper T cells. However, recent developments in molecular immunology have unequivocally demonstrated that the same gene products (α and β) make up the antigen-specific receptor on the two T cell types (DAVIS 1985; CHIEN et al. 1984; SAITO 1984). It is possible that the phenotypic difference between the two cells stems from the expression of a third chain (γ) present in mature CTL but not helpers (RAULET et al. 1985); however, this is not certain at the present time.

In addition to using the same germ line genes, the two T cell subsets are now thought to recognize structurally similar molecules on their partner cells (SCHWARTZ 1986). Class I molecules in association with $\beta 2$ microglobulin appear to possess a domain structure and 2-fold axis of symmetry similar to that of class II molecules (KAUFMAN et al. 1984). Thus, one would expect recognition of antigen in association with both MHC-encoded molecules to be similar. Yet for a long time it was felt that CTL recognized intact antigens on the surface of the target cell. Recently, however, it has been demonstrated in several systems that CTL are capable of recognizing fragments of the protein for which they are specific (WABUKE–BUNOTI et al. 1981; GOODING and O'CONNELL 1983; TOWNSEND et al. 1984). Thus, T-cell recognition of peptides in association with MHC-encoded molecules appears to be a universal mechanism for T-cell activation. Therefore, a careful evaluation of synthetic peptides as vaccines would seem to be a reasonable intellectual undertaking.

3 Peptides as Vaccines

The general strategy used to define a T-cell antigenic determinant in animal models has been to immunize a mouse or guinea pig with the foreign molecule in Freund's complete adjuvant and then to test the draining lymph node T-cell population for a proliferative response to both the intact protein and species variants of the molecule (SCHWARTZ 1985). Any evolutionary variants that give stimulation (or less informatively, fail to give stimulation) are examined to determine which residues are shared with the immunogen and which are different. This gives a clue as to which sites on the molecule might be important. The exact position of the determinant is identified by chemically or enzymatically cleaving the molecule and testing the fragments for stimulation of T cells primed to the whole molecule. The site is then confirmed by making a synthetic peptide of the suspected region and using this peptide to stimulate in vitro T cells primed to either the whole molecule or the peptide. Further refinements on the size of the determinant and the critical nature of each residue can then be carried out with a series of synthetic peptides of decreasing size and varied composition.

This approach has been used with a variety of protein antigens such as cytochrome *c*, lysozyme, myoglobin, and the influenza hemagglutinin. Presumably, it could be adapted to man by using natural infection as the method of priming and peripheral blood lymphocytes as the source of T cells for assay. Immunization with the synthetic peptides during clinical trials on normal volunteers could be used to confirm the correctness of the localization.

At this point, however, the first problem in the peptide vaccine approach begins to arise. Not all individuals will respond to any one antigenic determinant. As indicated above, the allelic form of the histocompatibility molecule(s) strongly influences which peptides are recognized by the immune system. Inbred strains of mice differing only in the MHC focus on different antigenic determinants. For example, in the helper/inducer T-cell response to lysozyme, mice possessing *k* alleles in the *I* region of the MHC respond predominantly to fragments 46–61 and 74–86, whereas mice possessing *b* alleles respond predominantly to fragment 81–96 and mice possessing *d* alleles respond to a determinant around residues 113 and 114 (GOODMAN et al. 1983; ALLEN et al. 1985). Thus, in order to develop a synthetic peptide vaccine applicable to an entire population, individuals with each major MHC haplotype would have to be studied to ascertain which peptides they predominantly use as T-cell determinants. Then a composite vaccine containing all the different peptides would have to be created.

Although this approach is intellectually feasible, it is by no means simple or likely to be inexpensive. It is estimated that anywhere from 50 to 100 different allelic forms of class II molecules exist in outbred murine populations (KLEIN and FIGUEROA 1981). The number in humans is probably greater since multiple loci encoding class II molecules are known to exist (KAUFMAN et al. 1984). If each allelic form of the Ia molecule functioned optimally with only one unique T-cell determinant, the work of identifying all the determinants and creating the corresponding synthetic peptides would be enormous. Even if there were some overlap, i.e., if certain peptides could function to some degree with several allelic forms of the class II molecule, and even though most humans are heterozygous at these genetic loci and their potential for interacting with any given peptide is therefore expanded, the result would likely be a suboptimal vaccine with different degrees of efficacy for different members of the population.

Furthermore, once one is forced into trying to create a universal vaccine by mixing a variety of synthetic peptides, other problems begin to emerge. Although poorly understood at the cellular level, the phenomenon of a suppressor feedback mechanism in the immune response is well documented (TADA and OKUMURA 1980; GERMAIN and BENACERRAF 1981). In certain cases it has been demonstrated that cells having suppressor activity are specific for unique antigenic determinants on the molecule, different from those recognized by the helper T cells (GOODMAN and SERCARZ 1983). For example, in the lysozyme system, residues at the amino terminal end of the molecule can elicit suppressor cells, whereas residues in the middle and at the carboxy terminal end of the molecule elicit helper and proliferative cells. The problem, however, is that elicitation of suppressor cells varies with the MHC haplotype of the mouse strain: $H\text{-}2^b$-bearing strains generate the suppressors, $H\text{-}2^k$-bearing strains do not. In fact, a suppressor determinant in one strain could turn out to be a helper determinant in another. Thus, a peptide that is useful for priming one individual in the population could end up inhibiting the response in another individual.

A similar problem may be introduced with regard to autoimmunity. If elicitation of an immune response by an infectious agent occasionally primes T cells that cross-react with self proteins in the context of that individual's MHC molecules, recognition of the self determinant in the tissues of the individual could produce inflammation and cell damage. This situation, like all T-cell responses, would be specific for a particular combination of peptide and MHC molecule. Thus, a helper determinant ideal for one individual in the population might elicit a strong autoimmune response in another individual. In order to circumvent these problems, it would appear that the ideal vaccine for any given individual would have to be custom made!

One final point that should be discussed is the use of peptides to generate T cell determinants which are normally not presented to the immune system. In some cases T-cell clones could exist that are specific for a given MHC molecule and peptide, but the peptide is not generated in sufficient amounts during antigen processing to stimulate these clones. A synthetic peptide vaccine could circumvent the need for antigen processing and thus allow such clones to be stimulated. This approach would be useful in situations where T-cell help is lacking and a primary antibody response is protective. However, a major problem arises if a secondary immune response is required. When the infectious organism subsequently invades, it must be processed in order to generate peptides that will activate the T cells. Because the determinant used to prime these particular T cells is inefficiently produced, they are unlikely to be reactivated. In such cases the vaccine would be useless.

4 Alternative Strategies

Many of the effective vaccines now available undoubtedly owe some of their efficacy to their ability to prime T cells. An examination of these vaccines suggests that a synthetic peptide approach may not be necessary. In general, two empirical approaches have been used. If the protein being administered has a negative biologic effect, such as diphtheria toxin, the molecule is chemically denatured (in this case, by formalin treatment). Although this procedure undoubtedly destroys a few T cell antigenic determinants, most remain intact. The effect on B cell recognition, which is more conformation sensitive, is probably much greater, but even some of these determinants survive. The processing of the denatured molecule might also be altered, causing the loss of still more T-cell antigenic determinants. Creating new determinants from the altered processing is of little value because they cannot be formed during subsequent infection and processing of the native protein. Despite all this potential destruction of T-cell antigenic determinants by denaturing the protein, numerous studies in experimental animals have shown that denatured molecules retain sufficient antigenic determinants to prime T cells reactive with native proteins in most animals and strains tested (KATZ 1977). Thus, this approach is clearly capable of circumventing the problem of *Ir* gene controlled nonresponsiveness stemming from poor antigen-Ia molecule interactions.

A second approach is used when the molecule to be sensitized against has no serious negative biologic consequences, e.g., the influenza hemagglutinin. In such cases the infectious organism is usually inactivated to prevent replication (e.g., by UV treatment) or attenuated in virulence by serial passage in vitro. The latter procedure most likely selects for inactivating mutations in proteins deleterious to the host. The net result is a vaccine containing the molecule in its native structure. This should be recognized by the immune system, both B and T cells, in exactly the same way as with natural infection. Thus, the only problem that could arise is the potential eliciting of suppressor T cells or an autoimmune reaction in some individuals in the population.

Given the documented successes of many current vaccines, it seems to me that the most straightforward strategy for designing a new vaccine is a recombinant DNA approach. The immunology of any system still has to be studied to determine which molecules of the infectious organism the immune system focuses on to produce an effective defense. Then the genes encoding these molecules could be isolated and expressed in *E. coli* or yeast to produce large amounts of the protein for immunization, or expressed in vaccinia virus (or other carriers) for direct vaccination. In those cases where the protein itself is toxic, site-directed mutagenesis could be used to inactivate the active site on the molecule without significantly destroying its immunogenicity.

This approach preserves the native structure of the antigen, insuring optimum priming of both T and B cells. It also provides all the T-cell antigenic determinants available on the molecule, thus insuring priming of T cells in most individuals. Finally, it might even be feasible to inactivate a few T-cell determinants by site-directed mutagenesis if they turned out to elicit suppressor T cells or autoimmune reactions in a large enough segment of the population being immunized.

In sum, I think a recombinant DNA approach offers a substantially better method for vaccine development than does a peptide approach. It is also likely to be more cost effective, because it would not require the immunological development costs needed to map out and confirm all the T-cell antigenic determinants on the molecules of interest. On the other hand, our current knowledge of T cell recognition is still quite primitive. In particular, which T cells become primed when multiple determinants on the molecule can be recognized, a phenomenon known as immunodominance, and whether this influences the efficacy of the immune response, remain to be explored. It is possible that peptides may play some role as vaccines if they can shift the balance in favor of T-cell clones that are more effective in providing a host defense. Finally, peptide vaccines may prove of some use in agriculture, e.g., for the immunization of farm animals, which are relatively inbred and thus might possess only limited polymorphism in the MHC.

Acknowledgements. I would like to thank Ronald Germain and John Weinstein for critically reviewing the manuscript and Shirley Starnes for expert editorial assistance.

References

Allen PM, Unanue ER (1984) Differential requirements for antigen processing by macrophages for lysozyme-specific T cell hybridomas. J Immunol 132:1077–1079

Allen PM, Matsueda GR, Haber E, Unanue ER (1985) Specificity of the T cell receptor: two different determinants are generated by the same peptide and the I-A^k molecule. J Immunol 135:368–373

Babbitt BP, Allen PM, Matsueda G, Haber E, Unanue ER (1985) Binding of immunogenic peptides to Ia histocompatibility molecules. Nature 317:359–361

Chien Y, Becker DM, Lindsten T, Okamura M, Cohen DI, Davis MM (1984) A third type of murine T cell receptor gene. Nature 312:31–35

Davis M (1985) Molecular genetics of T-cell receptor beta chain. Annu Rev Immunol 3:537–560

Geiger B, Rosenthal KL, Klein J, Zinkernagel RM, Singer SJ (1979) Selective and unidirectional membrane redistribution of an H-2 antigen with an antibody-clustered viral antigen: relationship to mechanisms of cytotoxic T-cell interactions. Proc Natl Acad Sci USA 76:4603–4607

Germain RN, Benacerraf B (1981) A single major pathway of T-lymphocyte interactions in antigen-specific immune suppression. Scand J Immunol 13:1–10

Gooding LR, O'Connell KA (1983) Recognition by cytotoxic T lymphocytes of cells expressing fragments of the SV 40 tumor antigen. J Immunol 131:2580–2586

Goodman JW, Sercarz EE (1983) The complexity of structures involved in T-cell activation. Annu Rev Immunol 1:465–498

Katz DH (1977) Determinant specificity of T and B cell receptors. In: Lymphocyte differentiation, recognition, and regulation. Academic, New York, pp 202–204

Kaufman JF, Auffray C, Korman AJ, Shackelford DA, Strominger J (1984) The class II molecules of the human and murine major histocompatibility complex. Cell 36:1–13

Klein J, Figueroa F (1981) Polymorphism of the mouse H-2 loci. Immunol Rev 60:23–57

Kovac Z, Schwartz RH (1985) The molecular basis of the requirement for antigen processing of pigeon cytochrome c prior to T cell activation. J Immunol 134:3233–3240

Laver WG, Air GM (1985) Immune recognition of protein antigens. Cold Spring Harbor Laboratory, New York, pp 1–197

Lerner RA (1984) Antibodies of predetermined specificity in biology and medicine. Adv Immunol 36:1–44

Raulet DH, Garman RD, Saito H, Tonegawa S (1985) Developmental regulation of T-cell receptor gene expression. Nature 314:103–107

Saito H, Kranz DM, Takagaki Y, Hayday AC, Eisen HN, Tonegawa S (1984) A third rearranged and expressed gene in a clone of cytotoxic T lymphocytes. Nature 312:36–40

Schwartz RH (1984) The role of gene products of the major histocompatibility complex in T cell activation and cellular interactions. In: Paul WE (ed) Fundamental immunology. Raven, New York, pp 379–438

Schwartz RH (1985) T-lymphocyte recognition of antigen in association with gene products of the major histocompatibility complex. Annu Rev Immunol 3:237–261

Schwartz RH (1986) Immune response (*Ir*) genes of the murine major histocompatibility complex. Adv Immunol 38:31–201

Shimonkevitz R, Kappler J, Marrack P, Grey HM (1983) Antigen recognition by H-2 restricted T cells. I. Cell-free antigen processing. J Exp Med 158:303–316

Streicher HZ, Berkower IJ, Busch M, Gurd FRN, Berzofsky JA (1984) Antigen conformation determines processing requirements for T-cell activation. Proc Natl Acad Sci USA 81:6831–6835

Tada T, Okumura K (1980) The role of antigen specific T cell factors in the immune response. Adv Immunol 28:1–87

Townsend ARM, McMichael AJ, Carter NP, Huddleston JA, Brownlee GG (1984) Cytotoxic T cell recognition of the influenza nucleoprotein and hemagglutinin expressed in transfected mouse L cells. Cell 39:13–25

Unanue ER (1984) Antigen presenting function of the macrophage. Annu Rev Immunol 2:395–428

Wabuke-Bunoti MAN, Fan DP, Braciale TJ (1981) Stimulation of anti-influenza cytolytic T lymphocytes by CNBr cleavage fragments of the viral hemagglutinin. J Immunol 127:1122–1125

Zinkernagel RM, Doherty PC (1979) MHC-restricted cytotoxic T cells: studies on the biological role of polymorphic major transplantation antigens determining T-cell restriction-specificity, function, and responsiveness. Adv Immunol 27:52–177